Study Guide

for

Moore and McCabe's

Introduction to the Practice of Statistics

Third Edition

William I. Notz
Michael A. Fligner
both of
Ohio State University

W. H. Freeman and Company
New York

Printed in the United States of America

ISBN: 0-7167-3400-1

Third Printing, 2001, Hamilton

CONTENTS

PREFACE TO THE STUDY GUIDE

We have written this study guide to help you learn and review the material in the text. The study guide is structured as follows. We first provide an overview of each section, which reviews the key concepts of that section. After the overview, there are guided solutions to selected odd-numbered problems in that section, along with the key concepts from the section required for solving each problem. The guided solutions provide hints for setting up and thinking about the exercise, which should help improve your problem-solving skills. Once you have worked through the guided solution, you can look up the complete solution provided later in the study guide to check your work.

What is the best way to use this study guide? Part of learning involves doing homework problems. In doing a problem, it is best for you to first try to solve the problem on your own. If you are having difficulties, then try to solve the problem using the hints and ideas in the guided solution. If you are still having trouble, then you can read through the complete solution. Generally, try to use the complete solution as a way to check your work. Be careful not to confuse reading the complete solutions with doing the problems themselves. This is the same mistake as reading a book about swimming and then believing you are prepared to jump into the deepest end of the pool. If you simply read our complete solutions and convince yourself that you could have worked the problem on your own, you may be misleading yourself and could have trouble on exams. When you are having difficulty with a particular type of problem, find similar problems to work in the exercises in the text. Often problems adjacent to each other in the text use the same ideas.

In the overviews we try to summarize the material that is most important, which should help you review material when you are preparing for a test. If any of the terms in the overview are unfamiliar, you probably need to go back to the text and reread the appropriate section. If you are having difficulties with the material, don't neglect to see your instructor for help. Face to face communication with your instructor is the best way to clear up difficulties.

We would like to thank Jackie Miller for reading an early draft of the study guide and providing helpful comments. The responsibility for any remaining errors is ours.

CHAPTER 1

EXAMINING DISTRIBUTIONS

SECTION 1.1

OVERVIEW

Section 1.1 introduces several methods for exploring data. These methods should be applied only after clearly understanding the background of the data collected. The choice of method depends to some extent upon the type of variable being measured. The two types of variables described in this section are

- **Categorical variables**: Variables that record to what group or category an individual belongs. Hair color and gender are examples of categorical variables. Although we might count the number of people in the group with brown hair, we wouldn't think about computing an average hair color for the group, even if numbers were used to represent the hair color categories.

- **Quantitative variables:** Variables that have numerical values and with which it makes sense to do arithmetic. Height, weight, and GPA are examples of quantitative variables. It makes sense to talk about the average height or GPA of a group of people.

To summarize the **distribution** of a variable, for categorical variables use **bar charts** or **pie charts**, and for numerical data use **histograms** or **stemplots**. Also, when numerical data are collected over time, a **timeplot** can be used in addition to a histogram or stemplot to look for interesting features of the data. When examining the data through graphs, we should be on the alert for

- Unusual values that do not follow the pattern of the rest of the data

- Some sense of a central or typical value of the data

- Some sense of how spread out or variable the data are

- Some sense of the shape of the overall pattern

In addition, when drawing a timeplot be on the lookout for **trends** occurring over time. Although many of the graphs and plots may be drawn by computer, it is still up to you to recognize and interpret the important features of the plots and the information they contain.

GUIDED SOLUTIONS

Exercise 1.3

KEY CONCEPTS - individuals and type of variables

Identify the "individuals" or objects described, then the "variables" or characteristics being measured. Once the variables are identified, you need to determine if they are categorical (the variable just puts individuals into one of several groups) or quantitative (the variable takes meaningful numerical values for which arithmetic operations make sense).

The "individuals" in this problem are the vehicles. If we included price, this would be a quantitative variable. If we had another variable, say a 1 if the car was available with four-wheel drive and 0 if it wasn't, this would still be a categorical variable even though we used numbers to represent the two categories. Now list the variables recorded and classify each as categorical or quantitative.

Name of variable   Type of variable

Exercise 1.7

KEY CONCEPTS - interpreting variables

To determine what a variable is telling us, we must know the purpose for which the variable is to be used. If the purpose is to make comparisons, as here, we need to consider whether the groups being compared differ only in the value of the variable or if they differ in other ways. If the groups differ in other ways, ask yourself if these differences could be partly responsible for differences in the value of the variable. For example, if two groups differ in size, then variables related to size (such as counts) are likely to differ even though the groups are identical in all other respects. In this exercise, consider ways in which the groups (different years) might differ and if these might explain the differences in cancer death rates even if cancer treatments are becoming more effective. Here are some scenarios that can be applied to different parts of the problem.

• Suppose that cancer was detected earlier. What would that do to survival rates?

• Suppose that population size increases and 1% of the population dies of cancer each year.

• Suppose that progress is made on the fight against heart disease. What would be the effect on the death rates due to cancer?

Exercise 1.17

KEY CONCEPTS - interpreting a histogram

Suppose that in some states the weaker students take the SATs and have a distribution with a single peak around 500. Sketch the histogram for this group. Now, suppose for the other states that a group of stronger students takes the SATs and has a distribution with a single peak around 550. Sketch the histogram for this group. What happens when you put both groups together in a single histogram?

Think of other situations for which a histogram with a shape similar to this problem might arise.

Exercise 1.23

KEY CONCEPTS - drawing histograms and stem-and-leaf plots and interpreting their shapes

DRP scores

40	26	39	14	42	18	25	43	46	27	19
47	19	26	35	34	15	44	40	38	31	46
52	25	35	35	33	29	34	41	49	28	52
47	35	48	22	33	41	51	27	14	54	45

How to draw a histogram:

1. When drawing a histogram, choose class intervals that divide the data into classes of equal length. For this data set, the smallest DRP score is 14 and the largest is 54, so the class intervals need to cover this entire range. A simple set of class intervals would be 10–<20 (10 is included in the interval but not 20), 20–<30, 30–<40, 40–<50 and 50–<60. Other sets of intervals are possible, although these have the advantage of using fairly simple numbers as endpoints.

2. Count the number of data values in each class interval. Using the class intervals above, complete the frequency table below. Remember, when counting the number in each interval, be sure to include data values equal to the lower endpoint but not the upper endpoint.

Interval   Count   Percent

3. Draw the histogram, which is a picture of the frequency table. Draw the bars and label the axes. Using the frequency table you have computed, complete the histogram given. The x-axis has been labeled for you. Be sure to include an appropriate label for the y-axis, as well as numbers for the scale.

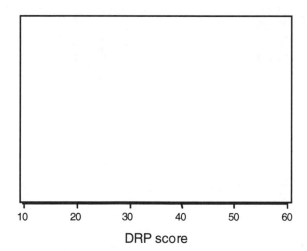

DRP score

How to draw the stem-and-leaf plot - single stems and splitting stems

1. It is easiest, although not necessary, to order the data first. If the data have been ordered, the leaves on the stems will be in increasing order. The DRP scores have been ordered for you below.

14 14 15 18 19 19 22 25 25 26 26 27 27 28 29 31 33 33 34 34
35 35 35 35 38 39 40 40 41 41 42 43 44 45 46 46 47 47 48 49
51 52 52 54

2. Using the stems below, complete the stem-and-leaf plot.

```
1 |
2 |
3 |
4 |
5 |
```

What are the similarities between this stem-and-leaf plot and the histogram you drew? Thinking about the relationship between the stems and the class intervals in this example, will help to explain why the stem-and-leaf plot looks just like the histogram laid on its side. What class intervals do the stems correspond to? What is one important difference between the histogram and the stem-and-leaf plot?

3. Using the DRP scores, increase the number of stems by splitting each of the previous stems in two. Complete the split stem-and-leaf plot in the space given.

In this case, splitting the stems results in a plot similar to a histogram, that uses too many class intervals. (How would you construct a histogram that corresponds to this stem-and-leaf plot? What would the class intervals be?) The more regular features of the data set are becoming obscured by the extra details you are forced to look at with the extra stems. As a display of the data, would you prefer the histogram or the first stem-and-leaf plot? Why?

To finish up the example, think about the important features that describe a distribution. Does the distribution of the DRP scores have a single peak? Does it appear to be symmetric, or is it skewed to the right (tail with larger values is longer) or to the left? Are there any outliers that fall outside the overall pattern of the data?

Exercise 1.35

KEY CONCEPTS - drawing and interpreting a timeplot

a) Complete the timeplot on the graph below. The first point is given.

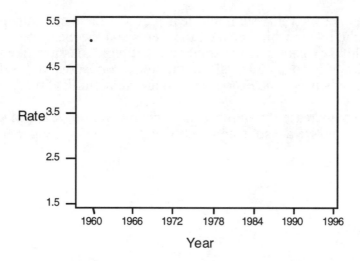

What is the general pattern in the timeplot? What happens to the time features of the data when a histogram or stem-and-leaf plot is drawn?

COMPLETE SOLUTIONS

Exercise 1.3

Name of variable	Type of variable
Type	Categorical
Where made	Categorical
City MPG	Quantitative
Highway MPG	Quantitative

Exercise 1.7

a) As the population size increases and people also live longer, the number of people dying each year from cancer, as well as from other causes, will go up even if treatments are more effective, simply because there are more people. For example, suppose the number of people at risk for cancer (because of age) increases from 1000 to 1500. Suppose treatments have reduced the incidence of cancer from 5% to 4%. The number of cancer deaths will increase—from 50 to 60!

b) People die of something—if other death rates went down, the cancer death rate could still go up even if cancer treatments were more effective. People are more likely to survive long enough to contract and die from cancer than from another disease.

c) Suppose that cancer was detected earlier, but treatments were not more effective. People would appear to live longer just as a result of earlier diagnoses. For example, if breast cancer was being diagnosed one year earlier but life expectancy was not changed by treatment, patients would appear to be surviving one year longer. (A woman with breast cancer dies at 55: she was diagnosed at 52 instead of 53, so her "survival" time is 3 years instead of 2.)

Exercise 1.17

Bimodal (two-peak distributions) often arise when the individuals on whom the variables are being measured form two distinct groups. Some other examples might be arm strength (men and women would have distributions centered about different values) or salaries at a university (staff and faculty would have distributions centered about different values). The next graph is a histogram with shaded bars for the first group and unshaded bars for the second group. When looking at the overall histogram, we wouldn't see these two pieces—we would see a single histogram, with bar heights equal to the sum of the shaded and unshaded heights for each interval. If we take the two individual histograms below and combine them, the resulting histogram will have two distinct peaks. Note that if the centers of the two histograms get too close, the two distinct modes tend to disappear.

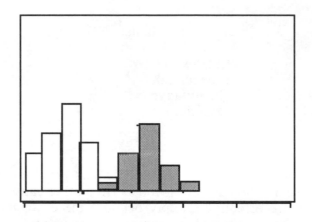

Exercise 1.23

Interval	Count	Percent
10 - < 20	6	13.63%
20 - < 0	9	20.45%
30 - < 40	11	25.00%
40 - < 50	14	31.82%
50 - < 60	4	9.09%
	44	99.99%

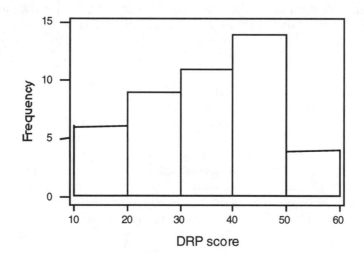

```
Stem-and-leaf plot of DRPscore, with single stems

1| 445899
2| 255667789
3| 13344555589
4| 00112345667789
5| 1224

Stem-and-leaf plot of DRPscore, with split stems

1| 44
1| 5899
2| 2
2| 55667789
3| 13344
3| 555589
4| 0011234
4| 5667789
5| 1224
```

In general, for this number of observations, the preference for a histogram over a stem-and-leaf plot is a personal preference. The stem-and-leaf plot is usually preferred for smaller data sets, the histogram for larger data sets. In this particular example, the data set is probably starting to get a little large for the stem-and-leaf plot, which, by keeping a record of every observation, is beginning to look a little cluttered. So our preference would be a histogram: it seems to make the general shape of the distribution a little more apparent.

Comments on the general shape: There are no obvious outliers that depart from the general pattern. The distribution is unimodal and skewed to the left. Test scores, which have an upper bound on the maximum score that students come close to, are often left-skewed. Although there is a lower bound of zero, the scores generally don't go quite that low, but instead slowly trail off on the lower end, giving the left-skewed appearance.

Exercise 1.35

a, b) There is a fairly steady downward trend in the timeplot from 1966 on. There is no particular change or noticeable feature that occurs around 1974 or in the 1980s. Don't read too much into the three identical rates from 1976 to 1980. The general trend is downward regardless of the speed limit, and some leveling off occurs in other places as well.

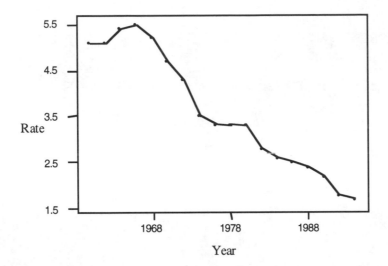

c) The interesting features in these data are the patterns over time. All of those features would be lost in a histogram. If the data were presented as a histogram, it would be impossible to answer the question in (b).

SECTION 1.2

OVERVIEW

Although graphs give an overall sense of the data, numerical summaries of features of the data make the notions of center and spread more precise.

Two important measures of center are the **mean** and the **median.** If there are n observations, $x_1, x_2,...,x_n$, then the mean is

$$\bar{x} = \frac{x_1 + x_2 + ... + x_n}{n} = \frac{1}{n}\sum x_i$$

where $\sum$ means "add up all these numbers." Thus, the mean is just the total of all the observations divided by the number of observations.

Although the median can be expressed by a formula, it is simpler to describe the rules for finding it.

How to find the median.

1. List all the observations from smallest to largest.

2. If the number of observations is odd, then the median is the middle observation. Count from the bottom of the list of ordered values up to the $(n + 1)/2$ largest observation. This observation is the median.

3. If the number of observations is even, then the median is the average of the two center observations.

The most important measures of spread are the **quartiles,** the **standard deviation**, and **variance.** For measures of spread, the quartiles are appropriate when the median is used as a measure of center. In general, the median and quartiles are more appropriate when outliers are present or when the data are skewed. In addition, the **five-number summary**, which reports the largest and smallest values of the data, the quartiles and the median, provides a compact description of the data that can be represented graphically by a **boxplot**. Computationally, the first quartile, Q_1, is the median of the lower half of the list of ordered observations and the third quartile, Q_3, is the median of the upper half of the list of ordered values.

If you use the mean as a measure of center, then the standard deviation and variance are the appropriate measures of spread. Remember that means and variances can be strongly affected by outliers and are harder to interpret for skewed data.

If we have n observations, $x_1, x_2, ...,x_n$, with mean $\bar{x}$, then the variance s^2 can be found using the formula

$$s^2 = \frac{(x_1 - \bar{x})^2 + (x_2 - \bar{x})^2 + ... + (x_n - \bar{x})^2}{n - 1} = \frac{1}{n-1}\sum(x_i - \bar{x})^2$$

The standard deviation is the square root of the variance, i.e., $s = \sqrt{s^2}$, and is a measure of spread in the same units as the original data. If the observations are in feet, then the standard deviation is in feet as well.

GUIDED SOLUTIONS

Exercise 1.43

KEY CONCEPTS - measures of center, five-number summary, drawing boxplots

a) Complete the back-to-back stemplot using the stems below. There are only a few observations over a fairly wide range, so the overall shapes of the distributions tend to be indistinct.

```
Women          Men
            7
            8
            9
           10
           11
           12
           13
           14
           15
           16
           17
           18
```

b) To find the means, find the sum of the scores, then divide by the number of scores. For the median, since the number of scores is even for both groups, the median is the average of the two middle scores. You can find this easily from the stemplot.

The feature that would suggest that $\bar{x} > M$ is not necessarily the same for both distributions, so look at them carefully.

c) The five-number summary consists of the median, the minimum and maximum, and the first and third quartiles. Remember that the first quartile is the median of those observations below the median (for the women, there are nine observations below the median) and the third quartile is the median of those observations above the median. Fill in the table below. These numbers are used to draw the boxplot.

	Men	Women
Minimum		
Q_1		
M		
Q_3		
Maximum		

To determine if the 1.5 × IQR criterion flags the largest women's observation, first compute the IQR. It is $Q_3 - Q_1$.

$$\text{IQR} =$$

$$1.5 \times \text{IQR} =$$

Then add 1.5 × IQR to the third quartile and see if the largest women's observation exceeds this value.

$$Q_3 + (1.5 \times \text{IQR}) =$$

Is 200 larger than this number? If so, then it is flagged as an outlier. We have drawn the boxplot for the men. After making sure you understand the men's boxplot, add the women's boxplot to this picture. First draw the box, which goes from the first quartile to the third quartile. Next, locate the median within the box. Finally, check for outliers. If there are no outliers, the lines from the box extend to the smallest and largest observations. If there are outliers, then the lines from the box extend to the smallest and largest observations that are not outliers. The outliers are then identified individually with a symbol (usually either a * or a dot).

Drawing one or two boxplots or other graphical displays by hand is the best way to make sure you understand how to interpret the display. But after that, it is really best to leave the drawing of boxplots and most other graphical displays to a statistical computer package.

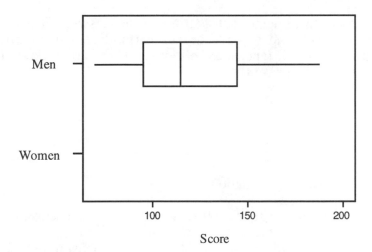

d) Use both the stem-and-leaf plot and the boxplot to answer the questions. Which graphic makes it easier to answer the questions?

Exercise 1.49

KEY CONCEPTS - measures of center

When there are several observations at a single value, the key is to remember that the mean is the total of all the observations divided by the number of observations. When computing the total, remember to include a salary as many times as it appears. The same is true when ordering the observations to find the median: remember to include a salary as many times as it appears.

Exercise 1.51

KEY CONCEPTS - measures of center, resistant measures

The change of extremes affects the mean but not the median. To compute the new mean, you can figure out the new total (you don't need to add up all the numbers again; just think about how much it has gone up) and divide by the number of observations. Or else you can think about dividing up the salary increases by the number of observations and adding this value to the old mean.

Exercise 1.55

KEY CONCEPTS - standard deviation

There are two points to remember in getting to the answer. The first is that numbers "further apart" tend to have higher variability than numbers closer together. The second is that repeats are allowed. There are several choices for the answer to (a) but only one for (b).

Exercise 1.65

KEY CONCEPTS - linear transformations

This is the first exercise in recognizing when a new measurement can be expressed in terms of an old measurement by the equation $x_{new} = a + bx$. This form of transforming an old measurement to a new measurement is called a linear transformation. In part (a) you need to convert horsepower to watts, and find the values of a and b for the transformation.

If one horsepower is 746 watts, then 2 horsepower would be 746 x 2 =1492 horsepower, and in general $x_{new} = 746x$. This corresponds to a linear transformation with $a = 0$, and $b = 746$. To finish the problem, a 140-horsepower engine would have $(140)(746) = 10440$ watts of power. Try and set up the other parts yourself. Linear transformations are important in statistics and will appear at several points in the book.

COMPLETE SOLUTIONS

Exercise 1.4.3
a)

Women		Men
	7	05
	8	8
	9	12
931	10	489
5	11	3455
966	12	6
77	13	2
80	14	06
442	15	1
55	16	9
8	17	
	18	07
	19	
0	20	

b) For the men, the sum of the 20 scores is 2425 and the mean is $\bar{x} = 121.25$. For the women, the sum of the 18 scores is 2539 and the mean is $\bar{x} = 141.06$. Since the number of scores is even for both groups, the median is the average of the two midddle scores. For the men M = 114.5 and for the women M = 138.5. For the women there is really little skewness - the mean exceeds the mean because of the outlier, and only slightly. For the men, the distribution is right-skewed, and the mean exceeds the median because of this.

c)

	Men	Women
Minimum	70	101
Q_1	98	126
M	114.5	138.5
Q_3	143	154
Maximum	187	200

The IQR for women is 154 − 126 = 28, and 1.5 × IQR = 37. If we add 37 to the third quartile, we get 154 + 37 = 191. Since 200 exceeds this, it is flagged as an outlier.

d) The boxplot makes the comparison of the plots easier. The symmetry of the women's scores, with the exception of the outlier, is fairly obvious, as is the right–skewness of the men's distribution. It is also somewhat clearer from examining the boxplot that the women's scores tend to be higher than the men's, whereas the men's are more variable.

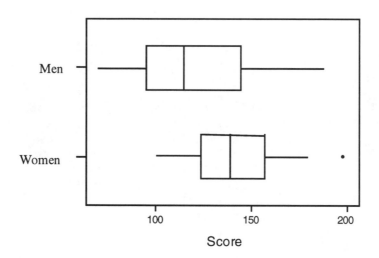

Exercise 1.49

The number of observations (individuals) is 5 + 2 + 1 = 8 and the total of the salaries is

$$(5 \times 25{,}000) + (2 \times \$60{,}000) + (1 \times \$255{,}000) = \$500{,}000$$

The mean is $500,000/8 = $62,500. Everyone earns less than the mean, except for the owner.

Since there are eight observations, the median is the average of the fourth and fifth smallest observations, which is $25,000. To see this, the eight ordered observations are

$25,000, $25,000, $25,000, $25,000, $25,000, $60,000, $60,000, $255,000

Exercise 1.51

The owner has an increase in salary from $255,000 to $455,000, or an increase of $200,000. The total is increased by this amount, from $500,000 to $700,000, and the new mean is $700,000/8 = $87,500. Another way of thinking about it is that the $200,000 increase averaged among the 8 people is $25,000, so the mean must go up $25,000 to $62,500 + $25,000 = $87,500.

The fourth and fifth smallest observations are still the same, so the median is unaffected.

Exercise 1.55

a) The standard deviation is always greater than or equal to zero. The only way it can equal zero is if all the numbers in the data set are the same. Since repeats are allowed, just choose all four numbers the same to make the standard deviation equal to zero. Examples are 1, 1, 1, 1 or 2, 2, 2, 2.

b) To make the standard deviation large, numbers at the extremes should be selected. So you want to put the four numbers at 0 or 10. The correct answer is 0, 0, 10, 10. You might have thought 0, 0, 0, 10 or 0, 10, 10, 10 would be just as good, but a computation of the standard deviation of these choices shows that two at either end is the best choice.

c) There are many choices for (a) but only one for (b).

Exercise 1.65

a) If 1 horsepower is 746 watts, then 2 horsepower would be 746 × 2 = 1492 watts, and in general $x_{new} = 746x$ ($a = 0$, and $b = 746$). So a 140-horsepower engine would have (140)(746) = 104,440 watts of power.

b) To change from miles to kilometers, the conversion is miles = 0.62 kilometers or kilometers = miles/0.62. In general $x_{new} = x/0.62 = 1.6129x$ (1/.062 = 1.6129) ($a = 0$ and $b = 1.6129$). So 65 miles per hour is (1.613)(65) = 104.84 kilometers per hour.

c) The difference between x and 98.6 is $x - 98.6$ (positive numbers correspond to pool temperatures above body temperature and negative numbers to pool temperatures below body temperature). In general, $x_{new} = (x - 98.6)$ ($a = -98.6$ and $b = 1$).

d) A food with 30 milligrams corresponds to 100% of the RDA, 15 milligrams to 50%, and 60 milligrams to 200%.

In general,

$$(\%RDA) = 100\frac{(\text{number of milligrams in food})}{30}$$

($a = 0$ and $b = 10/3$).

SECTION 1.3

OVERVIEW

Section 1.3 considers the use of mathematical models (mathematical formulas) to describe the overall pattern of a distribution. The name given to a mathematical model that summarizes the shape of a histogram is a **density curve**. The density curve is a kind of idealized histogram. The area under a density curve between two numbers represents the proportion of the data that lies between these two numbers. Like a histogram, it can be described by measures of center such as the **median** (a point such that half the area under the density curve is to the left of the point), the **mean** (the center of gravity or balance point of the density curve), and **modes** (major peaks in the density curve).

One of the most commonly used density curves in statistics is the **normal curve**. Normal curves are symmetric and bell-shaped. The peak of the curve is located above the mean and median, which are equal since the density curve is symmetric. The standard deviation measures how concentrated the area is around this peak. Normal curves follow the 68 – 95 – 99.7 rule; i.e., 68% of the area under a normal curve lies within one standard deviation of the mean (illustrated in the figure below), 95% within two standard deviations of the mean, and 99.7% within three standard deviations of the mean.

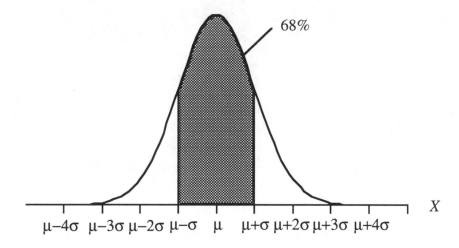

Areas under any normal curve can be found easily if quantities are first **standardized** by subtracting the mean from each value and dividing the result by the standard deviation. This standardized value is sometimes called the **z-score**. If data whose distribution can be described by a normal curve are standardized (all values replaced by their z-scores), the distribution of these standardized values is described by the **standard normal curve**. Areas under standard normal curves are easily computed by using a standard normal table such as that found in Table A in the front inside cover of the text.

If we know that the distribution of data is described by a normal curve, we can make statements about what values are likely and unlikely, without actually observing the individual values of the data. Although one can examine a histogram or stem-and-leaf plot to see if it is bell-shaped, the preferred method for determining if the distribution of data is described by a normal curve is a **normal quantile plot**. These plots are easily made using modern statistical computer software. If the distribution of data is described by a normal curve, the normal quantile plot should look like a straight line.

Normal
quantile
plot

In general, density curves are useful for describing distributions. Many statistical procedures are based on assumptions about the nature of the density curve that describes the distribution of a set of data. **Density estimation** refers to techniques for finding a density curve that describes a given set of data.

GUIDED SOLUTIONS

Exercise 1.69

KEY CONCEPTS - density curves and area under a density curve

Figure 1.35 is reproduced below. Although not indicated in the figure, the height of the density curve is 1.

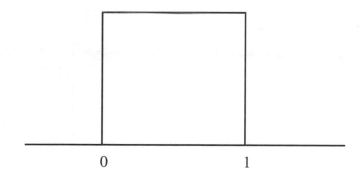

$$0 \qquad\qquad 1$$

a) The area between 0 and 1 under the density curve forms a square whose sides are of length 1. Recall that the area of any rectangle is the product of the length of the base of the rectangle and the height. In this case both are 1, so the area is 1.

b) The area of interest is the shaded region in the figure below.

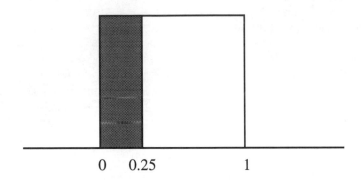

Compute the area of the shaded region by filling in the blanks below.

area = length of base × height = _____ × _____ = _____

c) Try answering this part on your own, using the same reasoning as in (b). First, shade in the area of interest in the figure below.

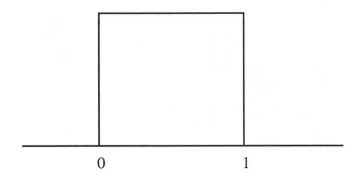

Now compute the area of your shaded region as in (b).

Exercise 1.77

KEY CONCEPTS - the 68 – 95 – 99.7 rule for normal density curves

Recall that the 68 – 95 – 99.7 rule says that the area under a normal curve between the mean minus one standard deviation and the mean plus one standard deviation is 0.68, the area under a normal curve between the mean minus two standard deviations and the mean plus two standard deviations is 0.95, and the area under a normal curve between the mean minus three standard deviations and the mean plus three standard deviations is 0.997. Also recall that the area under a density curve between two numbers corresponds to the proportion of data that lies between these two numbers.

In this problem, the mean is 266 days and the standard deviation is 16 days. From the 68 – 95 – 99.7 rule we have, for example, that the middle 68% of the lengths of all pregnancies lie between $266 - 1 \times 16 = 250$ and $266 + 1 \times 16 = 282$ days.

a) What does the 68 – 95 – 99.7 rule say about the middle 95% of the lengths of all pregnancies?

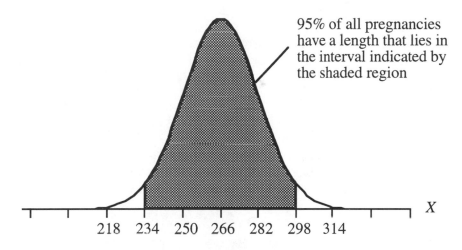

95% of all pregnancies have a length that lies in the interval indicated by the shaded region

218 234 250 266 282 298 314

We see from the figure that the middle 95% of all pregnancies have a length between 234 and 298 days.

b) We indicated above that the middle 68% of all pregnancies have lengths between 250 and 282 days. Thus 100% – 68% = 32% are either below 250 or above 282. What percentage must be below 250 (recall that the density curve is symmetric)? What percentage must be above 282?

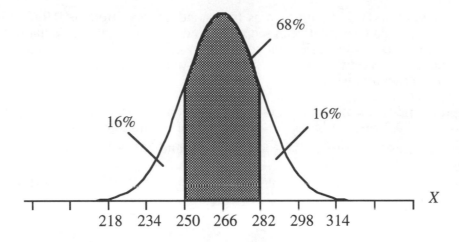

Now use this reasoning with the middle 95%.

Exercise 1.81

KEY CONCEPTS - computing areas under a standard normal density curve

Recall that the proportion of observations from a standard normal distribution that are less than a given value z is equal to the area under the standard normal curve to the left of z. Table A gives these areas, illustrated in the figure below.

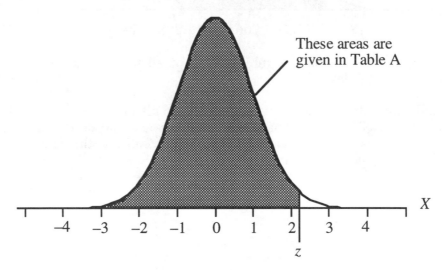

In answering questions about the proportion of observations from a standard normal distribution that satisfies some relation, we find it helpful to first draw a picture of the area under a normal curve corresponding to the relation. We then try to visualize this area as a combination of areas of the form in the previous figure, since such areas can be found in Table A. The entries in Table A are then combined to give the area corresponding to the relation of interest.

This approach is illustrated in the solutions that follow.

a) A picture of the desired area is given below.

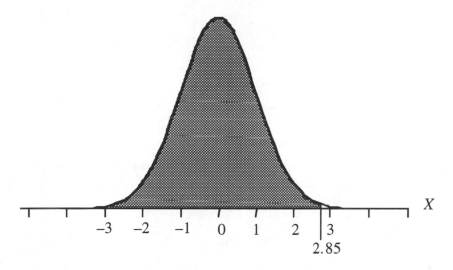

This is exactly the type of area that is given in Table A. Simply find the row labeled 2.8 along the left margin of the table, locate the column labeled 0.05 across the top of the table, and read the entry at the intersection of this row and column. This entry is 0.9978. This is the proportion of observations from a standard normal distribution that satisfies $Z < 2.85$.

b) Shade in the desired area in the picture below.

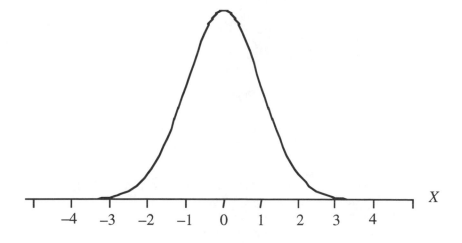

Remembering that the area under the curve is 1, how would you modify your answer from (a)?

c) Try solving this part on your own. To begin, draw a picture of a normal curve and shade the region.

Now use the same line of reasoning as in (b) to determine the area of your shaded region. Remember, you want to try to visualize your shaded region as a combination of areas of the form given in Table A.

d) To test yourself, try this part on your own. It is a bit more complicated than the previous parts, but the same approach will work.

Exercise 1.83

KEY CONCEPTS - finding the value z (the quantile) corresponding to a given area under a standard normal density curve

The strategy used to solve this type of problem is the "reverse" of that used to solve Exercise 1.81. We again begin by drawing a picture of what we know; we know the area, but not *z*. For areas corresponding to those given in Table A we have a situation like the following.

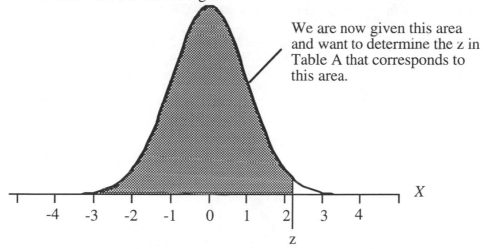

We are now given this area and want to determine the z in Table A that corresponds to this area.

To determine *z*, we find the given value of the area in the body of Table A (or the entry in Table A closest to the given value of the area). We now look in the left margin of the table and across the top of the table to determine the value of *z* that corresponds to this area.

If we are given a more complicated area, we draw a picture and then determine from properties of the normal curve the area to the left of *z*. We then determine *z* as described above. The approach is illustrated in the solutions below.

a)   A picture of what we know is given below. Note that since the area given is less than 0.5, we know *z* must be to the left of 0 (recall that the area to the left of 0 under a standard normal curve is 0.5).

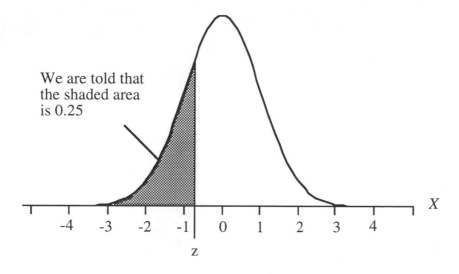

We are told that the shaded area is 0.25

We now turn to Table A and find the entry closest to 0.25. This entry is 0.2514. Locating the z values in the left margin and top column corresponding to this entry, we see that the z that would give this area is 0.67.

b) Try this part on your own. Begin by sketching a normal curve and the area you are given on the curve. On which side of zero should z be located? Thinking about which side of zero a point lies on is a good way to make sure your answer makes sense.

Exercise 1.87

KEY CONCEPTS - computing the area under an arbitrary normal density curve

For these problems, we must first convert the question into one about a standard normal. This involves standardizing the numerical conditions by subtracting the mean and dividing the result by the standard deviation. We then draw a picture of the desired area corresponding to these standardized conditions and compute the area as we did for the standard normal, using Table A. This approach is illustrated in the solutions below.

a) We need to first standardize the condition $X > 700$. We replace X by Z (we use Z to represent the standardized version of X) and standardize 700. Since we are told that the mean and standard deviation of GRE scores are 544 and 103, respectively, the standardized value (z-score) of 700 is (rounded to two decimal places)

$$z\text{-score of } 700 = (700 - 544)/103 = 1.51$$

Our condition in "standardized" form is

CONDITION: $Z > 1.51$.

A picture of the desired area is

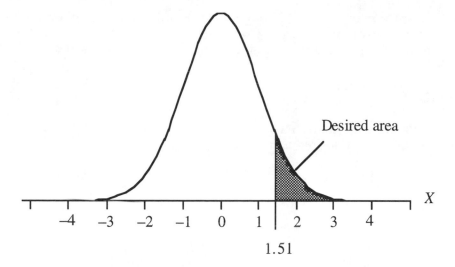

The desired area is not of the form given in Table A. However we note that the unshaded area to the left of 1.51 is of the form given in Table A and this area is 0.9345. Hence

shaded area   = total area under normal curve – unshaded area
            = 1 – 0.9345 = 0.0655

Thus, the proportion of applicants whose score X satisfies $X > 700$ is 0.0655.

b) Try this part on your own. First compute z-scores to convert the problem to a statement involving standardized values. In terms of z-scores, the condition of interest is

CONDITION:

Sketch the standard normal curve below and shade the desired region on your curve.

Use Table A to compute the desired area. This will be the answer to the question.

c) Try this part on your own. Although this is a little more complicated than (b), use the same line of reasoning. Convert values to z-scores to express the condition in terms of standardized values. In terms of z-scores, the condition of interest is

CONDITION:

Sketch the standard normal curve below, shade the desired area, and then use Table A to compute the desired area.

Exercise 1.94

KEY CONCEPTS - finding the value x (the quantile) corresponding to a given area under an arbitrary normal density curve

To solve this problem, we must use an approach that is the reverse of that used in Exercise 1.87. Conceptually, we think of having standardized the problem. We then find the value z for the standard normal distribution that satisfies the stated condition; i.e., has the desired area. We next must "unstandardize" this z value by multiplying by the standard deviation and then adding the mean to the result. This unstandardized value x is the desired result. We illustrate this strategy in the solutions below.

We are told in Exercise 1.93 that the WISC scores are normally distributed with μ = 100 and σ = 15. To find the score that will place a child in the top 5% of the population, we first find the corresponding value z for the standard normal. This value z must have the property that the area to the right of it under the standard normal curve is 0.05. This is illustrated in the figure below.

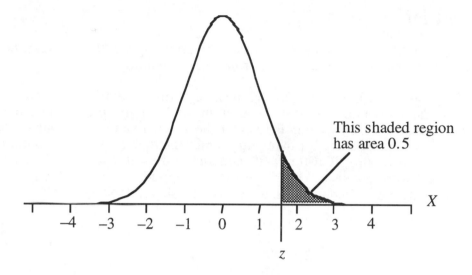

This shaded region has area 0.5

From the figure, we see that the area to the left of z (the unshaded area) is

unshaded area = total area under normal curve − shaded area = $1 - 0.05 = 0.95$

These are the types of areas reported in Table A. Find the entry in the body of Table A that has a value closest to 0.95. This entry is 0.9495. The value of z that yields this area is seen, from Table A, to be 1.64.

We now must unstandardize z. The unstandardized value is

x = (standard deviation) $\times z$ + mean = $15z + 100 = 15 \times 1.64 + 100 = 124.6$

Thus, a child must score at least 124.6 to be in the top 5%. Assuming that fractional scores are not possible, a child would have to score at least $x = 125$ to place in the top 5%.

Now see if you can determine the score needed for a child to place in the top 1%. Use the same line of reasoning as above. Begin by drawing the region representing the z value corresponding to the top 1%.

Exercise 1.101

KEY CONCEPTS - normal quanitle plots and determining whether the distribution of a set of data can be described by a normal density curve

Table 1.4 is given in Exercise 1.26. To make a normal quantile plot, you should use statistical software. Consult the user manual for the procedure for your software package. The first step is to enter the 73 data values. If your software package uses a spreadsheet for data entry, enter the data in a single column. If you have access to an ASCII (text) file containing the data, you should import it into your software package. Then use the appropriate command for making a normal quantile plot. Below is such a plot. Yours should look similar.

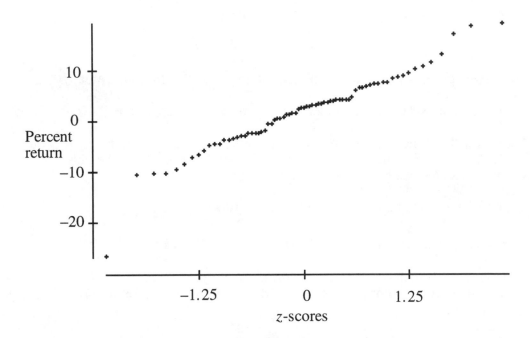

To interpret the plot, ask yourself the following questions:

  • Do the points appear to follow a straight line?
  • If not, in what ways do they deviate? Are there outliers? Are there any unusual "bends" at either end of the plot?

Refer to Figures 1.30 to 1.33 in your text for some guidance in interpreting your plot. You might also make a histogram of the data to check your interpretation. Write down your interpretation and check your answer with the solution provided.

COMPLETE SOLUTIONS

Exercise 1.69

a) A complete solution is given in the guided solution.

b) The area of interest is the shaded region indicated below.

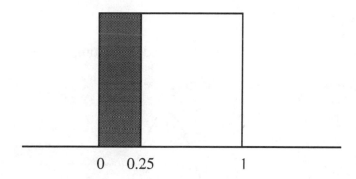

This rectangular region has area = length of base × height = $0.25 \times 1 = 0.25$.

c) The area of interest is the shaded region indicated below.

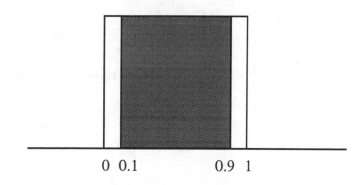

This rectangular region has area = length of base × height = $0.8 \times 1 = 0.8$.

Exercise 1.77

a) A complete solution is given in the guided solution.

b) Refer to the figure in (a) of the guided solution. If the shaded region gives the middle 95% of the area, then the two unshaded regions must account for the remaining 5%. Since the normal curve is symmetric, each of the two unshaded regions must have the same area and each must account for half of the 5%. Hence, each of the unshaded regions accounts for 2.5% of the area. The leftmost of these regions accounts for the shortest 2.5% of all pregnancies, and the rightmost accounts for the longest 2.5% of all pregnancies. We conclude

(i) The shortest 2.5% of all pregnancies have lengths of less than 234 days.

(ii) The longest 2.5% of all pregnancies have lengths of more than 298 days.

Exercise 1.81

a) A complete solution is given in the guided solution.

b) A picture of the desired area is given below.

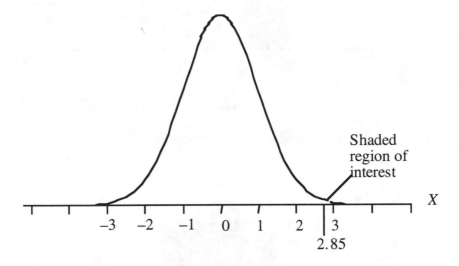

Table A cannot be used directly for this area. However, the unshaded area to the left of 2.85 is of the form needed for Table A. In fact, we found the area of the unshaded portion in (a). We notice that the shaded area can be visualized as that part left after deleting the unshaded area from the total area under the normal curve.

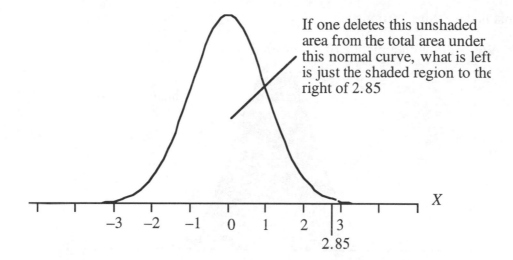

If one deletes this unshaded area from the total area under this normal curve, what is left is just the shaded region to the right of 2.85

Since the total area under a normal curve is 1, we have

shaded area = total area under normal curve − area of unshaded portion
= 1 − 0.9978 = 0.0022

Thus, 0.0022 is the desired proportion.

c) Again, we begin with a picture of the desired area.

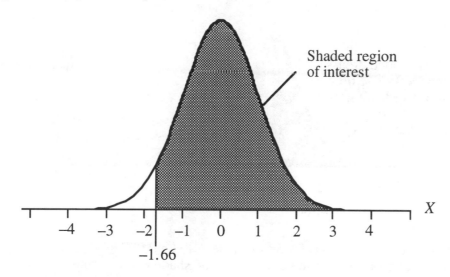

Shaded region of interest

This is just like (b). The unshaded area to the left of −1.66 can be found in Table A and is 0.0485. Thus,

shaded area = total area under normal curve − area of unshaded portion
= 1 − 0.0485 = 0.9515

This is the desired proportion.

d) We begin with a picture of the desired area.

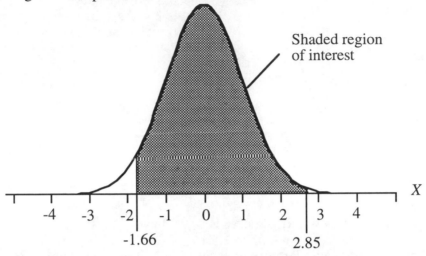

The shaded region is a bit more complicated than in the previous parts, however the same strategy still works. We note that the shaded region is obtained by removing the area to the left of -1.66 from the area to the left of 2.85.

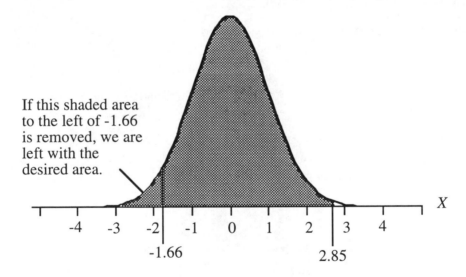

The area to the left of 2.85 is found in Table A to be 0.9978. The area to the left of -1.66 is found in Table A to be 0.0485. The shaded area is thus

shaded area = area to left of 2.85 - area to left of -1.66
                    = 0.9978 - 0.0485 = 0.9493

This is the desired proportion.

Exercise 1.83

a) A complete solution was given in the guided solution.

b) A picture of what we know is given below. Note that since the area to the right of 0 under a standard normal curve is 0.5, we know that z must be located to the right of 0.

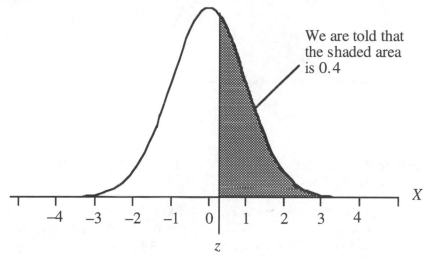

We are told that the shaded area is 0.4

The shaded area is not of the form used in Table A. However, we note that the unshaded area to the left of z is of the correct form. Since the total area under a normal curve is 1, this unshaded area must be $1 - 0.4 = 0.6$. Hence z has the property that the area to the left of z must be 0.6. We locate the entry in Table A closest to 0.6. This entry is 0.5987. The z corresponding to this entry is 0.25.

Exercise 1.87

a) A complete solution was given in the guided solution.

b) We need to first standardize the condition $X < 500$. We replace X by Z (we use Z to represent the standardized version of X) and standardize 500. Since we are told that the mean and standard deviation of GRE scores are 544 and 103, respectively, the standardized value (z-score) of 700 is (rounded to two decimal places)

$$z\text{-score of } 500 = (500 - 544)/103 = -0.43$$

Our condition in "standardized" form is $Z < -0.43$. A picture of the desired area is

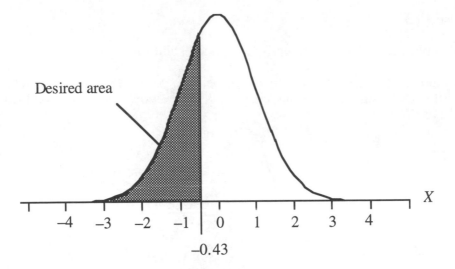

Desired area

−0.43

This area can be found directly in Table A and is 0.3336. Thus, the proportion of applicants whose score X satisfies $X < 500$ is 0.3336.

c) We need to first standardize the condition $500 < X < 800$. We replace X by Z (we use Z to represent the standardized version of X) and standardize 500 and 800. Since we are told that the mean and standard deviation of GRE scores are 544 and 103, respectively, the standardized value (z-score) of 700 is (rounded to two decimal places)

$$z\text{-score of } 500 = (500 - 544)/103 = -0.43$$

$$z\text{-score of } 800 = (800 - 544)/103 = 2.49$$

Our condition in "standardized" form is $-0.43 < Z < 2.49$. A picture of the desired area is

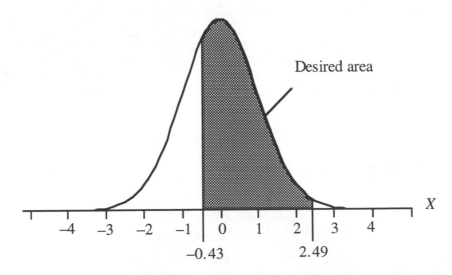

Desired area

−0.43   2.49

The desired area can be visualized as the area to the left of 2.49 with that portion to the left of –0.43 removed. From Table A, the area to the left of 2.49 is 0.9936, and the area to the left of –0.43 is 0.3336. Hence, the shaded area is

shaded area   = area to left of 2.49 – area to left of –0.43
           = 0.9936 – 0.3336 = 0.6600

Thus, the proportion of applicants whose score X satisfies $500 < X < 800$ is 0.6600.

Exercise 1.94

The complete solution for the top 5% is given in the guided solution.

To find the score that will place a child in the top 1% of the population, we first find the corresponding value z for the standard normal. This value z must have the property that the area to the right of it under the standard normal curve is 0.01. This is illustrated in the figure below.

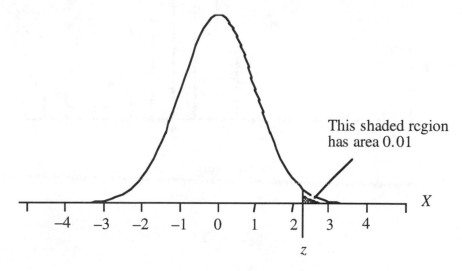

From the figure, we see that the area to the left of z (the unshaded area) is

unshaded area = total area under normal curve – shaded area = 1 - 0.01 = 0.99

These are the types of areas reported in Table A. We find the entry in the body of Table A that has value closest to 0.99. This entry is 0.9901. The value of z that yields this area is seen, from Table A, to be 2.33.

We now must unstandardize z. The "unstandardized" value is

x = (standard deviation) $\times z$ + mean = $15z + 100 = 15 \times 2.33 + 100 = 134.95$

Thus, a child must score at least 134.95 to be in the top 1%. Assuming that fractional scores are not possible, a child would have to score at least $x = 135$ to place in the top 1%.

Exercise 1.101

If the data follow a normal distribution, the plot should look approximately like a straight line. In our plot, the point in the lower left corner is an outlier. Ignoring this point, the rest of the plot is reasonably straight, suggesting that the remaining data are approximately normal. The very slight flattening of the left portion of the plot may indicate some "bunching" of the smallest values (ignoring the outlier). Making a histogram of the data, one sees these patterns.

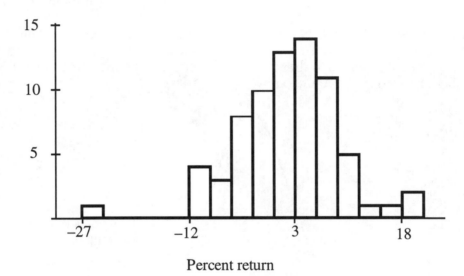

Percent return

CHAPTER 2

LOOKING AT DATA
RELATIONSHIPS

SECTION 2.1

OVERVIEW

The first chapter provides the tools to explore several types of variables one by one, but in most instances the data of interest are a collection of variables that may exhibit some kinds of relationships among themselves. Typically, these relationships are more interesting than the behavior of the variables individually. The first tool we consider for examining the relationship between variables is the **scatterplot**. Scatterplots show us two quantitative variables at a time, such as the weight of a car and its MPG (miles per gallon). Using colors or different symbols, we can add information to the plot about a third variable that is categorical in nature. For example, if we wanted to distinguish between cars with manual and automatic transmissions, we might use a circle to plot the cars with manual transmissions and a cross to plot the cars with automatic transmissions.

When drawing a scatterplot, we need to pick one variable to be on the horizontal axis and the other to be on the vertical axis. When there is a **response variable** and an **explanatory variable**, the explanatory variable is always placed on the horizontal axis. In cases where there is no explanatory-response variable distinction, either variable can go on the horizontal axis. After we draw the scatterplot by hand or using a computer, we should examine it for an **overall pattern** that may tell us about any relationship between the variables and, about **deviations** from it. We should be looking for the **direction**, **form**, and **strength** of the overall pattern. In terms of direction, **positive association** occurs both variables take on high values together, whereas **negative association** occurs if one variable takes high values when the other takes on low values. In many cases, when an association is present, the variables appear to have a **linear relationship**. The plotted values seem to form a line. If the line slopes up to the right, the association is positive; if the line slopes down to the right, the association is negative. As always, look for **outliers**. The outlier may be far away in terms of the horizontal variable or the vertical variable, or it may be far away from the overall pattern of the relationship.

GUIDED SOLUTIONS

Exercise 2.1

KEY CONCEPTS - explanatory and response variables

When examining the relationship between two variables, if you hope to show that one of the variables can be used to explain variation in the other, remember that the response variable measures the outcome of the study, while the explanatory variable explains or causes changes in the response variable. When you just want to explore the relationship between two variables, such as scores on the math and verbal SATs then the explanatory-response variable distinction is not important.

In (a), the grade on the exam is the response and the time spent studying is the explanatory variable. Now try the other parts on your own.

b)

c)

d)

e)

Exercise 2.5

KEY CONCEPTS - drawing and interpreting a scatterplot

a) When drawing a scatterplot, we first need to pick one variable (the explanatory variable) to be on the horizontal axis and the other (the response) to be on the vertical axis. In this data set, we are interested in the "effect" of drinking moderate amounts of wine on yearly deaths from heart disease. So wine consumption is the explanatory variable and deaths from heart disease is the response. We have drawn the points corresponding to Australia and Austria in the following plot. Although you will generally draw scatterplots on the computer, drawing a small one like this by hand ensures that you understand what the points represent.

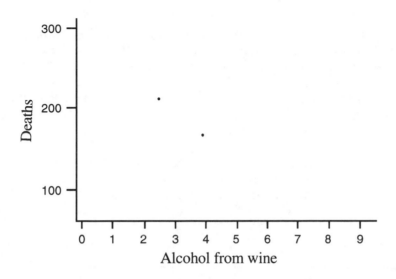

b) We are looking for the form and strength of the relationship. Can the relationship be described with a straight line? Section 2.3 discusses formal methods for drawing a straight line through a set of data, but for now just try to draw a straight line to follow the pattern in the scatterplot in (a). Do the points seem to follow the line that you have drawn? How tight is the scatter about that line?

c) Is the association positive or negative? Do countries with higher wine consumption tend to have higher or lower death rates? You need to be careful with the language you use to describe the relationship. In this example, the countries may differ on many other factors besides wine consumption, which may explain the lower death rates due to heart attacks. So avoid the use of expressions such as "drinking more wine lowers the risk of heart disease" or "wine produces a lower risk of heart disease," as these expressions imply causation. Try to explain in simple language what the data have to say.

Exercise 2.11

KEY CONCEPTS - drawing and interpreting a scatterplot with a categorical variable

a) Below is the graph of incubation period and age. In this example, the explanatory variable would be age. Why? What is its "effect" on the incubation period? The circles in the plot correspond to those who survived and the pluses to those who died. Most software that will draw a scatterplot will allow you to use different symbols for the points based on a third variable—all three variables are entered as part of the data set. If you are using a software package for this course, try to use it to make a plot similar to the one below.

Note: There are two deaths at age 32 with an incubation period of 48 hours, so there are only 17 distinct points on the graph instead of 18. Find the plus in the graph corresponding to these two equal observations.

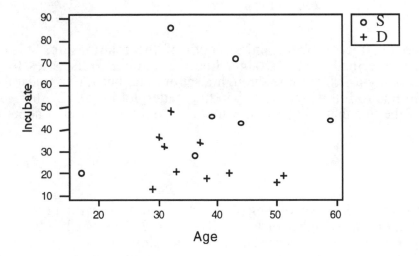

b) Ignoring the distinction between circles and pluses in the plot, is there an overall relationship between age and incubation period?

c) There are several things to look for here. Is there a different relationship between age and incubation for those who died and those who survived? Are the survivors different from those who died in terms of incubation period or age—that is, did the survivors tend to be younger or have longer incubation periods?

d) Unusual observations can be in the *x*-variable or *y*-variable, or they can be any observations that don't follow the general pattern of the data. There isn't much of a pattern in the data, so just look for unusual observations in terms of age or incubation. In which variable would it be more interesting to find an unusual observation?

Exercise 2.15

KEY CONCEPTS - categorical variables in scatter plots

a) In the scatterplot, there should be four observations above each of the levels of nematode count. After adding these points to the graph below, compute the mean of the four observations at each level of nematode count, and put each mean on the graph. Then connect the four means.

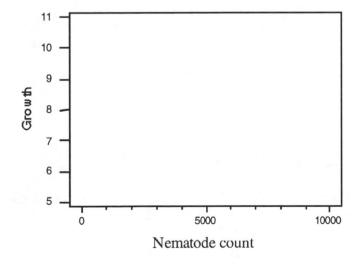

b) This question requires an informal comparison of the means. What sorts of changes do you see in the means as the nematode count increases?

COMPLETE SOLUTIONS

Exercise 2.1

b) The distinction is not important.

c) The response is the yield of a crop, and the explanatory variable is yearly rainfall.

d) The distinction not important.

e) The response is the son's occupational class and the explanatory variable is the father's occupational class. We believe that the father's occupational class has an "effect" on the son's and not the other way around. So there is a distinction between the two variables.

Exercise 2.5

a), b)

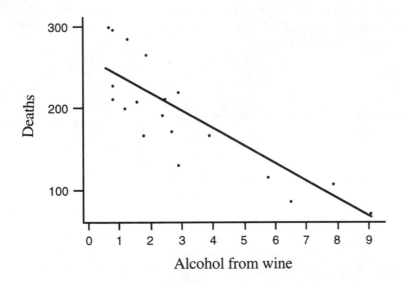

A line does not do a bad job of describing the general pattern. The relationship is moderately strong. In Section 2.2, we will give a numerical measure that describes the strength of the linear relationship.

c) There is a negative association between wine consumption and deaths from heart disease. Those countries in which wine consumption is higher tend to have a lower rate of deaths from heart disease. (Note: There is nothing in this language that implies causation.)

Exercise 2.11

a) The graph is given in the guided solution.

b) There doesn't appear to be an overall relationship between age and incubation.

c) If you examine just the circles or just the pluses, there doesn't appear to be a relationship between age and incubation for those who died or those who survived. The thing that stands out in the scatterplot is that survival appears to be related to the incubation period. If the incubation period was longer, the person seems more likely to have survived regardless of age. For the incubation periods of 21 hours or less, 6 out of 7 people died, while for incubation periods exceeding 21 hours, there are only 5 deaths out of 11 people (remember the double point if counting from the graph).

d) There isn't much of a pattern in the data, so just look for unusual observations in terms of age or incubation. There is one person who is younger than most (17 years), but that doesn't suggest much. There are two people with incubation periods that are much longer than the rest (72 and 86 hours), and we might want to examine those points a little further. Is there something that occurred differently with respect to those two individuals, possibly with respect to their exposure?

Exercise 2.15

a) In the following graph, the circles correspond to the observations and the pluses to the means at each level of nematode count.

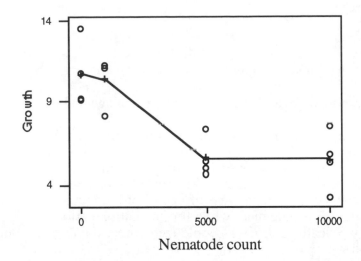

b) The level of growth may decrease slightly when going from 0 to a count of 1000, but then it drops off considerably by 5000 and seems to stay at that level until 10,000. In making these statements, we are making some assumptions about the average growth between the levels of nematode counts in our experiments (interpolating). We claimed that the growth level wasn't changing between 5000 and 10,000, but we have no data between those values. It is more problematic when the drop in level occurs to get from the level at 1000 to the level at 5000. It would be helpful to have had an observation at 3000, as the data suggest that there was a fairly sharp decrease in growth at some nematode count between 1000 and 5000, but we can't say more.

SECTION 2.2

OVERVIEW

Scatterplots provide a visual tool for looking at the relationship between two variables. Unfortunately, our eyes are not good tools for judging the strength of the relationship. Changes in the scale or the amount of white space in the graph can easily change our judgment of the strength of the relationship. **Correlation** is a numerical measure we will use to show the strength of **linear association**.

The correlation can be calculated using the formula

$$r = \frac{1}{n-1} \sum (\frac{x_i - \bar{x}}{s_x})(\frac{y_i - \bar{y}}{s_y})$$

where $\bar{x}$ and $\bar{y}$ are the respective means for the two variables X and Y, and s_X and s_Y are their respective standard deviations. In practice, you will probably be computing the value of r using computer software or a calculator that finds r from entering the values of the x's and y's. When computing a correlation coefficient, there is no need to distinguish between the explanatory and response variables, even in cases where this distinction exists. The value of r will not change if we switch x and y.

When r is positive, it means that there is a positive linear association between the variables, and when it is negative, there is a negative linear association. The value of r is always between 1 and –1. Values close to 1 or –1 show a strong association whereas values near 0 show a weak association. As with means and standard deviations, the value of r is strongly affected by outliers. The presence of outliers can make the correlation much different from what it might be with the outlier removed. Finally, remember that the correlation is a measure of straight line association. There are many other types of association between two variables, but these patterns will not be captured by the correlation coefficient.

GUIDED SOLUTIONS

Exercise 2.19

KEY CONCEPTS - interpreting and computing the correlation coefficient

a)

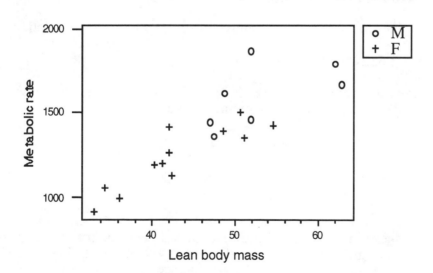

Should the sign of the correlation coefficient be the same for men and women? Is either relationship "stronger"? Are there outliers in either group that might raise or lower the value of the correlation coefficient?

b) Try and use a computer package or a calculator to compute the value of the correlation coefficient. If you do not have access to a calculator or computer package, the required "hand" computations are illustrated below for the men.

$$\bar{x} = 53.10, \qquad\qquad s_x = 6.69$$

$$\bar{y} = 1600.00, \qquad\qquad s_y = 189.2$$

We summarize the calculations for the correlation r in the following table

x	$\dfrac{x - \bar{x}}{s_x}$	y	$\left(\dfrac{y - \bar{y}}{s_y}\right)$	$\left(\dfrac{x - \bar{x}}{s_x}\right)\left(\dfrac{y - \bar{y}}{s_y}\right)$
62.0	1.33034	1792	1.01480	1.35003
62.9	1.46487	1666	0.34884	0.51100
47.4	-0.85202	1362	-1.25793	1.07178
48.7	-0.65770	1614	0.07400	-0.04867
51.9	-0.17937	1460	-0.73996	0.13273
51.9	-0.17937	1867	1.41121	-0.25313
46.9	-0.92676	1439	-0.85095	0.78862

The sum of the values in the last column above is 3.5524. Thus the correlation is

$$r = 3.5524/6 = 0.592 \text{ for the men.}$$

Now you need to either repeat the above "hand" calculation for the 12 women, or learn how to do the calculation on a computer package or calculator.

Women's correlation coefficient =

c) Mean body mass for men =
Mean body mass for women =

This is a difficult question. If the relationship remained the same for all weights between 40 and 65 kilograms (the range of all the data), then it wouldn't matter. The fact that the men were heavier wouldn't influence the correlation. If the relationship was different for heavier people than lighter people, then the fact that men are heavier could affect the value of the correlation because then men and women might have a different relationship between body mass and metabolic rate (with possibly differing strengths as measured by the correlation coefficient).

d) Is this a linear transformation? What is the effect of a linear transformation on the correlation coefficient?

Exercise 2.21

KEY CONCEPTS - computing the correlation coefficient, the effect of outliers

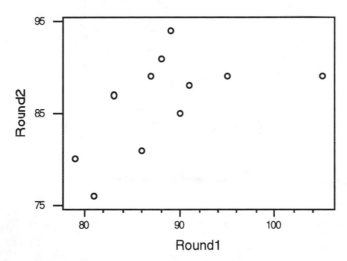

Player 7 clearly breaks the pattern of bad first round followed by a good second round. This point doesn't follow the linear relationship suggested by the other points, and including it weakens the overall linear relationship. This in turn lowers the numerical value of the correlation coefficient. Do the computations with and without this point.

Correlation with all observations   =
Correlation without observation 7   =

Exercise 2.27

KEY CONCEPTS - effect of transformations on the correlation coefficient

a) The first plot is speed (km/h) *vs.* fuel (liters/100 km) and the second is speed (miles/hour) *vs.* fuel (gallons/mile). Both are virtually identical except for the labeling of the axes, which corresponds to the units that the variables are measured in.

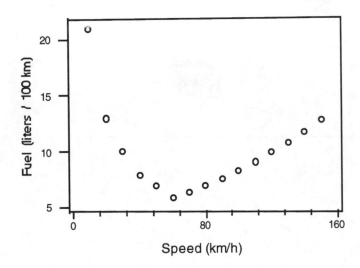

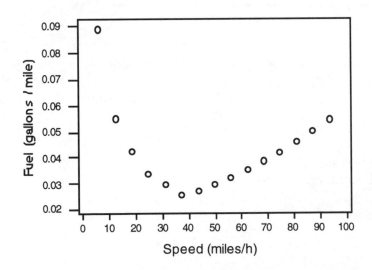

The transformation to change kilometers to miles is miles = kilometers/1.609 and is a linear transformation. What's the transformation to change fuel used in liters/100 km to fuel used in gallons/mile? Is it a linear transformation? What is the effect of these two transformations on the numerical value of the correlation coefficient?

Note: Since the relationship between speed and fuel consumption is not linear, do you think the correlation coefficient is good summary of the "strength" of the relationship?

b)

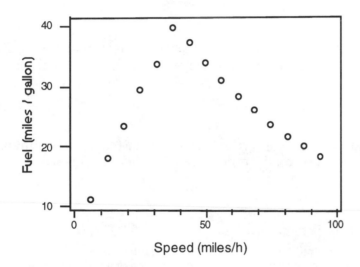

What is the transformation to change fuel used in gallons/mile to miles per gallon? What is the transformation to change liters/100 km to miles per gallon? Is this transformation linear? What is the effect on the correlation coefficient?

Note: The relationship is still not linear, so the correlation coefficient is again not a very good summary measure.

Exercise 2.31

KEY CONCEPTS - interpreting the correlation coefficient

The problem is that a correlation close to zero and the quote "good researchers tend to be poor teachers, and vice versa" are not the same. What does a correlation close to zero mean? What would be true about the correlation if "good researchers tend to be poor teachers, and vice versa"?

COMPLETE SOLUTIONS

Exercise 2.19

a) The relationship for the women seems to be tighter around a line. The men's observation of mass = 51.9 and rate = 1867 lowers the value of the correlation for the men.

b)   Men's correlation coefficient = 0.592
        Women's correlation coefficient = 0.876

c)   Mean body mass for men = 53.10.
        Mean body mass for women = 43.03

d) Kilograms = 2.2 × pounds is a linear transformation with $a = 0$ and $b = 2.2$ (see page 56). The value of the correlation is unchanged under linear transformations of the variables.

Exercise 2.21

The correlation with all 11 observations is 0.550. When player 7 is eliminated from the data set, the correlation on the remaining 10 observations is 0.661.

Exercise 2.27

a) The transformation here is fuel$_{(\text{gallons per mile})}$ = (1.609/378.5) × fuel$_{(\text{liters per 100 km})}$ and is linear. Since the transformation of x and y are both linear, the correlation coefficient is the same for the original and transformed variables. The value of the correlation coefficient is −0.172. Since the relationship is clearly nonlinear, and the correlation only measures the strength of a linear relationship, it would be a mistake to interpret this as a negative relationship between fuel consumption and speed.

b) To go from gallons per mile to miles per gallon we need to take the reciprocal miles per gallon = 1/gallons per mile, or

$$\text{mpg} = 378.5 \, / \, (1.609 \times \text{fuel}_{(\text{liters per 100 km})})$$

This is not a linear transformation, so the numerical value of the correlation coefficient will change. The value of the correlation coefficient is now −0.0430. A value near zero would suggest a weak relationship but that refers to a weak **linear** relationship. There is clearly a very strong relationship, between speed and miles per gallon, albeit nonlinear.

Exercise 2.31

If the correlation were close to zero, there would be no particular linear relationship. Good researchers would be just as likely as bad researchers to be good or bad teachers. The statement that "good researchers tend to be poor teachers, and vice versa" implies that the correlation is negative, not zero.

SECTION 2.3

OVERVIEW

If a scatterplot shows a linear relationship that is moderately strong as measured by the correlation, we would like to draw a line on the scatterplot to summarize the relationship. In the case where there is a response and an explanatory variable, the **least-squares regression** line often provides a good summary of this relationship. A straight line relating y to x has the form $y = a + bx$ where b is the **slope** of the line and a is the **intercept**. The least-squares regression line is the straight line $\hat{y} = a + bx$, which minimizes the sum of the squares of the vertical distances between the line and the observed values of y. The formula for the slope of the least squares line is

$$b = r\frac{s_y}{s_x}.$$

and for the intercept is $a = \bar{y} - b\bar{x}$, where $\bar{x}$ and $\bar{y}$ are the means of the x and y variables, s_x and s_y are their respective standard deviations, and r is the value of the correlation coefficient. Typically, the equation of the least-squares regression line is obtained by computer software or a calculator with a regression function.

Regression can be used to predict the value of y for any value of x. Just substitute the value of x into the equation of the least squares regression line to get the predicted value for y. Predicting values of y for x values in the range of those x's we observed is called interpolation and is fine to do. However, be careful about extrapolation. (using the line for prediction beyond the range of x values covered by the data). Extrapolation may lead to misleading results if the pattern found in the range of the data does not continue outside the range.

Correlation and regression are clearly related, as can be seen from the equation for the slope, b. However, the more important connection is how r^2, the square of the correlation coefficient, measures the strength of the regression. r^2 tells us the fraction of the variation in y that is explained by the regression of y on x. The closer r^2 is to 1, the better the regression describes the connection between x and y.

GUIDED SOLUTIONS

Exercise 2.35

KEY CONCEPTS - review of straight lines

a) To give the equation of a straight line, $y = a + bx$, the first thing to figure out is which variable will play the role of x and which will be y. In this problem,

$x =$

$y =$

The only thing remaining is to determine which piece of information represents the intercept or value of y when $x = 0$, and which represents the slope, how much y increases with each unit increase in x. In this problem,

$a =$

$b =$

The equation of the line is then

b) Draw the graph on the axes below. Remember, drawing a straight line requires only that you find two points on the line and connect them. The value at zero is easy, so you just need to pick a second value.

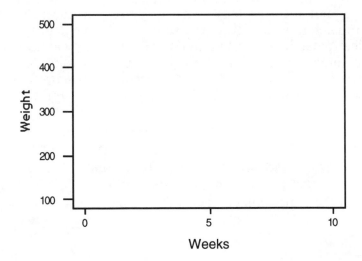

c) When predicting the weight at 2 years, remember the units in which age is measured before doing the calculation. Do you think that a rat's growth pattern will remain the same outside of the range of data? Why or why not?

Exercise 2.39

KEY CONCEPTS - scatterplots, r and r^2, least square regression, extrapolation

a) In drawing the scatterplot, we are thinking of boats as the explanatory variable and manatees killed as the response. You might also consider drawing a timeplot as in Chapter 1 to look at the data. There we would plot year *vs*. manatees. Because we believe that the important thing changing over time that is affecting the death rate is the number of boats, the plot given here is more meaningful, although it is not the only way to look at this data set. Given the scatterplot below, describe the form and direction of the relationship.

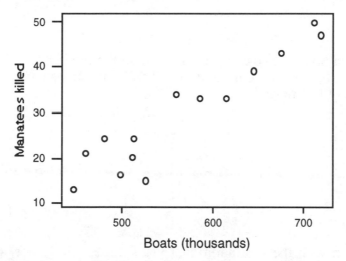

b) The correlation coefficient is a good way to summarize the strength of the relationship. Use a calculator or computer software to compute the correlation coefficient. How does the correlation coefficient relate to the fraction of variation explained by the number of boats registered? What does this say about the accuracy of the predictions?

c) Computing a regression line by hand can be quite tedious. We first illustrate the required calculations and then reproduce a portion of the output of the regression program using MINITAB.

$$\bar{x} = 567.50, \qquad\qquad s_X = 91.907$$

$$\bar{y} = 29.429, \qquad\qquad s_y = 12.189$$

$$r = 0.941$$

We calculate

$$b = r(s_y/s_X) = 0.941(12.189/91.907) = 0.1248$$

$$a = \bar{y} - b\bar{x} = 29.429 - 0.1248(567.50) = -41.395.$$

Regression Analysis

```
The regression equation is
Killed = - 41.4 + 0.125 Boats
```

Use this equation to predict the number of manatees that will be killed in a year when 716,000 powerboats are registered.

d) Convert 2 million boats into the correct units to do the prediction. What is the range of x-values that were used to fit the line? What can be said about predictions outside of this range?

e) Below is a graph with the four additional years added; these new points are plotted with a plus sign. When answering the question, consider the time order in which these four points were collected.

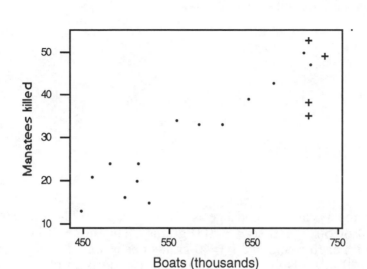

f) First compute the mean number of manatee deaths from 1991–1993 and then compare it to your prediction from part (c).

Exercise 2.49

KEY CONCEPTS - units of measurement for descriptive measures, effect of transformations on summary measures

a) The relevant portion of the MINITAB output for predicting metabolic rate based on lean body mass is given below.

Regression Analysis

```
The regression equation is
rate = 113 + 26.9 mass
```

What are the units of the slope, and what is it telling you?

b) $\bar{x}$ = 46.74, $\bar{y}$ = 1369.5, s_x = 8.28, s_y = 257.5, r = 0.865, b = 26.9, a = 113. These computations can be done easily using software. What are the units of measurement for each of these descriptive measures?

c) You need to figure which of these descriptive measures would have new units of measurement, a new value and what the effect of the linear transformation pounds=kilograms x 2.2 has on each quantity.

COMPLETE SOLUTIONS

Exercise 2.35

a) The weight is the y variable and age is the x variable. The intercept is the weight at age = 0 or birth, and so $a = 100$ grams. The slope is how much the weight goes up with each week, and $b = 40$ grams per week. The equation of the line is then $y = 100$ $40x$. Don't forget to include the units in the slope and the intercept.

b) At zero, the weight is equal to 100 grams, and at 10 weeks the value of weight is $100 + 40(10) = 500$ grams. Just plot these two points on the graph and connect them to get the line.

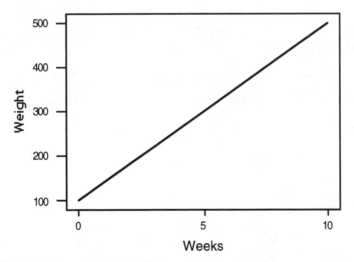

c) This is an example of extrapolation. Rats do not continue to grow at the same rate for 2 years, just as people wouldn't necessarily grow at the same rate for 20 years. At 2 years or 104 weeks (remember that x is measured in weeks), the predicted weight is $100 + 40(104) = 4260$ grams, which would be quite a rat!

Exercise 2.39

a) The form of the relationship looks linear, and the direction indicates a positive association.

b) Although the scatterplot suggests a moderately strong relationship, the correlation coefficient is a good way to quantify it. The numerical value of the correlation coefficient is 0.941. The fraction of the variation in manatee deaths explained by the number of boats registered is r^2, the square of the correlation coefficient, and is $(0.941)^2 = 0.89$. Thus, a substantial amount of the variation can be explained. The points are clustered relatively tightly about a line, suggesting that predictions over this range of years will be fairly accurate.

c) The regression equation is Killed $= -41.4 + 0.125$ Boats. Plug in the value 716 for boats (boats were measured in thousands of boats when fitting the line), and the prediction is $-41.4 + .125(716) = 48.1$, or approximately 48 manatees killed.

d) 2 million boats is 2000 thousand boats so the prediction is $-41.4 + 0.125(2000)$ $= 208.6$, or approximately 209 manatees. The range of values of x used to fit the line were from about 450 to 720. The value 2000 is well beyond this range, and this would be extrapolation.

e) Of the four new data points, two would suggest that the measures were helping. The number of manatees killed is well under what would have been predicted by the model. Unfortunately, in 1994, the last year for which data were collected, the number of deaths was back within the general pattern of the data before the measures were enacted. So we would have to say that things are inconclusive. More data are needed to see whether the trend toward lower deaths observed in 1992 and 1993 continues or as 1994 suggests, the stronger measures had only a temporary effect.

f) The mean for these three years was 42 and the prediction was 48.1. So for these three years, the death rate was under what was expected. Of course, the prediction assumed that conditions did not change. If measures were enacted to protect manatees, then predictions based on previous years would not apply.

Exercise 2.49

a) The regression equation is rate $= 113 + 26.9$ mass. The slope is 26.9 and the units are calories burned in 24 hours per kilogram of lean body weight. Over the range of the data (body weights between about 30 and 65 kilograms), every 1-kilogram increase in lean body mass is associated with an increase of about 26.9 calories burned in a 24-hour period.

b) The units of measurement are $\bar{x} =$ kilograms, $\bar{y} =$ calories, $s_x =$ kilograms, s_y $=$ calories, $r =$ unitless, $b =$ calories per kilogram, $a =$ calories.

c) $\bar{x} = 46.74 \times 2.2 = 102.83$ pounds, $\bar{y} = 1369.5$ calories (unchanged), $\qquad s_x$

$= 8.28 \times 2.2 = 18.22$ pounds, $s_y = 257.5$ calories (unchanged), $r = 0.865$ (unitless and unchanged), $b = 26.9/2.2 = 12.23$ calories per pound, $a = 113$ calories (unchanged). The regression equation is rate $= 113 + 12.23$ mass, where mass is now measured in pounds .

SECTION 2.4

OVERVIEW

Plots of the **residuals**, which are the differences between the observed and predicted values of the response variable, are very useful for examining the fit of a regression line. Features to look out for in a residual plot are unusually large values of the residuals (outliers), nonlinear patterns, and uneven variation about the horizontal line through zero (corresponding to uneven variation about the regression line).

The effects of **lurking variables**, variables other than the explanatory variable that may also affect the response, can often be seen by plotting the residuals versus such variables. Linear or nonlinear trends in such a plot are evidence of a lurking variable. If the time order of the observations is known, it is good practice to plot the residuals versus time order to see if time can be considered a lurking variable.

Influential observations are individual points whose removal would cause a substantial change in the regression line. Influential observations are often outliers in the horizontal direction.

Correlation and regression must be interpreted with caution. Plots of the data, including residual plots, help make sure the relationship is roughly linear and help to detect outliers and influential observations. The presence of lurking variables can make a correlation or regression misleading. Always remember that association, even strong association, does not imply a cause-and-effect relationship between two variables.

A correlation based on averages is usually higher than if we had data for individuals. A correlation based on data with a restricted range is often lower than would be the case if we had observed the full range of the variables.

GUIDED SOLUTIONS

Exercise 2.55

KEY CONCEPTS - residuals

a) If you are using statistical software, you should enter the data and use the software to create a scatterplot. Although we are giving you many of the plots, it is a good idea to make sure you understand in each plot why one variable was designated the response and the other the explanatory variable.

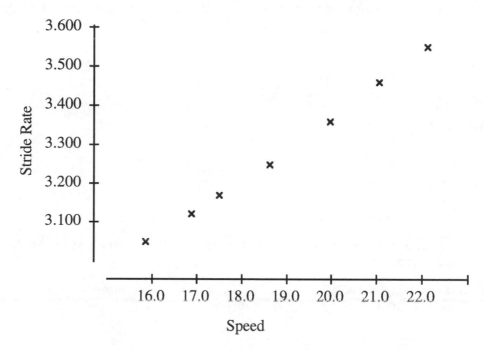

What is the general trend in your scatterplot? Does it appear to be adequately described by a straight line, or is curvature present?

b) If you do not have access to a computer or a calculator that will compute the least-squares regression line, you will have to do computations by hand (refer to Exercise 2.39). If you are using statistical software (or a calculator that will compute the equation of the least-squares regression line), you should enter the data and use the software (calculator) to calculate the equation of the least-squares regression line. Write the equation in the space provided below.

c) Recall that the residual for a given speed x is

$$\text{residual} = \text{observed stride rate} - \text{predicted stride rate}$$
$$\text{predicted stride rate} = a + bx$$

and $a + bx$ is the equation of the least-squares regression line from (b). Statistical software can be used to calculate the residuals directly. If you use statistical software, fill in the values in the residual column only in the table below. If you are calculating the residuals by hand, complete the table below to aid you in systematically calculating the residuals.

speed	observed stride rate	predicted stride rate	residual = observed - predicted
15.86	3.05		
16.88	3.12		
17.50	3.17		
18.62	3.25		
19.97	3.36		
21.06	3.46		
22.11	3.55		

Add the entries in the residual column to verify that the residuals sum to 0 (except for rounding error).

d) If you are using statistical software, the software should allow you to create a plot of the residuals directly. Plot the seven residuals on the axes below.

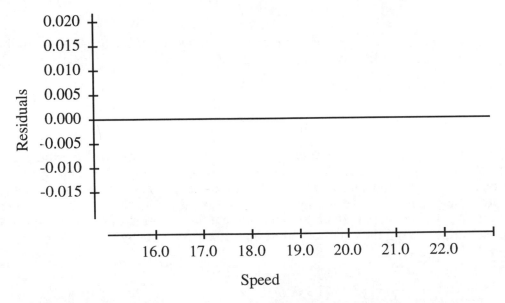

Does it appear that the residuals have a random scatter, or is there a pattern present? Do you think that a linear fit is appropriate for these data?

Do any of the points in your plot appear to be influential? Ask yourself if their removal would cause a substantial change in the least-squares line (or if they appear to be outliers in the horizontal direction).

Note: For classes that discussed the topic of DFFITS, you can also calculate the DFFITS to determine if any observations are influential.

Are you provided with any information that would allow you to plot the observations against the time they were made? You might ask yourself in what form this information would have to be in order to make such a plot (think about what the actual data values represent).

Exercise 2.65

KEY CONCEPTS - regression including a categorical variable, lurking variable

a) A plot of the data is given below. The x corresponds to the right-hand times and the circles to the left-hand times.

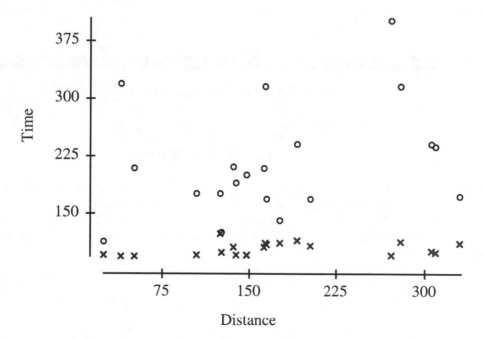

b) In describing the pattern, look for the most striking details. Are there any clear trends for the left-hand observations? The right-hand observations? Are there any differences between the left-hand and right-hand observations?

c) The calculations of the least-squares regression lines of time on distance for each hand are best done using statistical software. Be sure to analyze the left-hand and right-hand observations separately. If you must do them by hand, approach the calculations systematically as in the guided solution in Exercise 2.39. Write the two equations of the regression lines in the space provided below.

regression line for left-hand data:

regression line for right-hand data:

Now draw these lines carefully on the scatterplot in (a).

Which line appears to do the better job of predicting time from distance? You should compute the correlation r (or better yet, r^2) associated with each regression line as a possible numerical summary describing the success of the two lines. Can you reconcile the values of your numerical summaries with a visual inspection of the plot?

d) If you are plotting manually, you would first calculate the residuals using

$$\text{residual} = \text{time} - [99.364 + 0.028(\text{distance})]$$

for the right-hand observations and

$$\text{residual} = \text{time} - [171.548 + 0.262(\text{distance})]$$

for the left-hand observations. Then you would plot your results, remembering that the vertical axis represents the values of the residuals and the horizontal axis represents time. If you are using statistical software, your plots of the residuals from each regression against the time order of the trials should look similar to those given here.

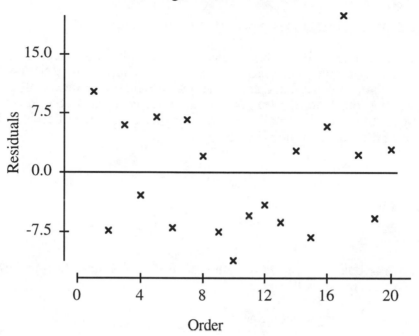

Plot for the right-hand observations

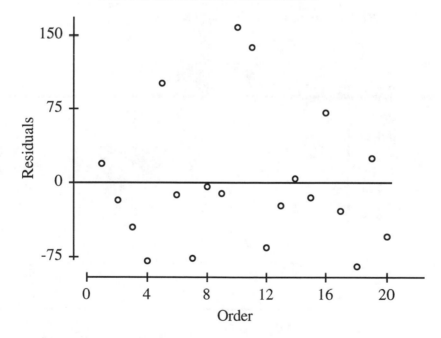

Plot for the left-hand observations

Are any trends or patterns visible in these plots?

Exercise 2.67

KEY CONCEPTS - correlations based on averaged data

Computing the correlation is most easily done using statistical software or a calculator that computes correlation. Regarding whether the correlation would increase or decrease if we had data on the individual stride rates of all 21 runners, the following quote from the text is relevant: "A correlation based on averages over many individuals is usually higher than the correlation between the same variables based on data for individuals."

COMPLETE SOLUTIONS

Exercise 2.55

a) A plot of the data is given in the guided solution. A straight line "appears" to describe the data quite well.

b) The least-squares regression line

$$\text{stride rate} = 1.77 + 0.08(\text{speed})$$

A scatterplot with this line drawn in is given below.

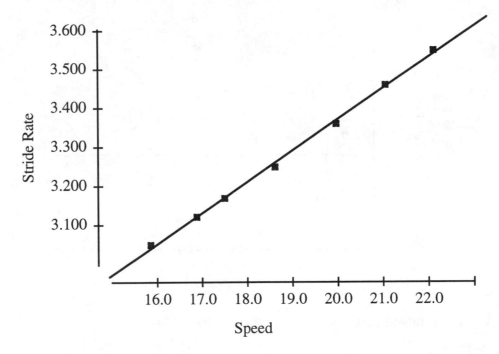

c) For each of the speeds given, we substitute the value into the equation of the least-squares regression line to compute the predicted value of stride rate. The residual is then computed as

residual = observed stride rate − predicted stride rate

For example, for a speed of x = 15.86, we compute

predicted stride rate = 1.77 + 0.08(15.86) = 1.77 + 1.27 = 3.04

and hence

residual = observed stride rate − predicted stride rate = 3.05 − 3.04 = 0.01

We summarize the results in the table below (to two decimal places)

speed	observed stride rate	predicted stride rate	residual
15.86	3.05	1.77 + 0.08(15.86) = 3.04	0.01
16.88	3.12	1.77 + 0.08(16.88) = 3.12	0.00
17.50	3.17	1.77 + 0.08(17.50) = 3.17	0.00
18.62	3.25	1.77 + 0.08(18.62) = 3.26	−0.01
19.97	3.36	1.77 + 0.08(19.97) = 3.37	−0.01
21.06	3.46	1.77 + 0.08(21.06) = 3.46	0.00
22.11	3.55	1.77 + 0.08(22.11) = 3.54	0.01

Although these calculations can be done by hand, they are much more easily obtained using statistical software and should agree (to two decimal places) with the above. If you sum the entries in the residual column, you can easily see that they sum to 0.

d) A plot of the residuals against speed is given on the following page

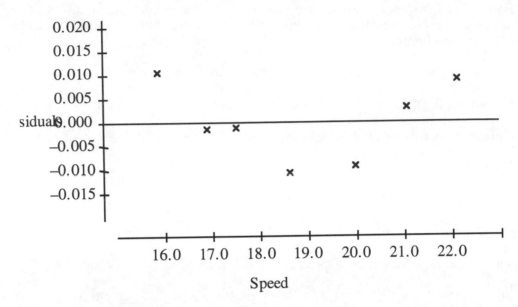

The pattern is curved (U-shaped) and indicates that the linear fit is not completely adequate. Recall that the scatterplot in (a) suggested that the linear fit was quite good. This demonstrates that the residual plot can be more informative than the scatterplot in determining the adequacy of the linear fit.

There do not appear to be any influential observations in the plot. Since we are not told the time at which observations were made, we cannot plot the residuals against the time at which observations were made. Note that since observations are actually the averages for 21 runners, we would have to know that all observations on the 21 runners at a given speed were made at the same time to even be able to make such a plot.

Note: For classes that discussed DFFITS, the values of DFFITS are given below. These are most easily computed using statistical software.

speed	stride rate	DFFITS
15.86	3.05	1.6635023
16.88	3.12	−0.08839246
17.50	3.17	−0.05792908
18.62	3.25	−0.59243516
19.97	3.36	−0.55913409
21.06	3.46	0.24486537
22.11	3.55	1.4562468

The first and last entries listed are the largest values of DFFITS and hence the most influential observations. Neither point appears to be particularly influential, however.

Exercise 2.65

a) The plot is given in the guided solution.

b) There does not appear to be any clear relation between distance and time. The most striking feature of the plot is that nearly all the left-hand observations (the O's) lie above the right-hand observations (the X's). This means that left-hand trials had longer times for a given distance than right-hand trials. The subject performed consistently better with the right hand, suggesting that the subject is right-handed.

c) Least-squares regression for the right-hand observations.

$$time = 99.364 + 0.028(distance)$$

Least-squares regression for the left-hand observations.

$$time = 171.548 + 0.262(distance)$$

These lines have been drawn on our scatterplot, reproduced below.

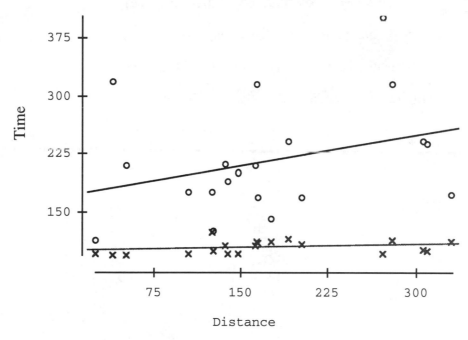

Visually, the regression line for the right-hand data appears to do a better job of predicting time from distance, since the X's appear to be more tightly clustered about this line than the O's are about the regression line for the left-hand data. This would suggest that the value of r^2 would be much higher for the right-hand data. The regression for the right-hand observations has $r^2 = 9.3\%$; i.e., the least-squares regression of time on distance explains 9.3% of the variation in the time values. The regression for the left-hand observations has $r^2 = 10.1\%$, i.e., the least-squares regression of time on distance explains 10.1% of the variation in the time values. Thus, the regression for the left-hand observations actually explains a higher percentage of the variation in time than does the regression for the right-hand observations.

For the right-hand data, there is little variation in times to explain. The times are all quite close to the mean value of 104.25, and this mean is a pretty good predictor of times *without* using distance as an explanatory variable (this is why the least-squares regression line is almost horizontal and why r^2 is so low. Distance adds little information to help with predictions). For the left-hand data, the r^2 value agrees with our visual impression of the plot. There is a lot of variability, and distance explains little, resulting in a low r^2.

d) From the equations of the two least-squares regression lines, we can calculate the residuals using the general expression

$$\text{residual} = \text{observed value of time} - \text{predicted value of time}$$

For the right-hand observations, this becomes

$$\text{residual} = \text{time} - [99.364 + 0.028(\text{distance})]$$

For the left-hand observations, this becomes

$$\text{residual} = \text{time} - [171.548 + 0.262(\text{distance})]$$

The following tables summarize the results of the above calculations.

Order	residuals(right-hand)	residuals(left-hand)
1	10.2369110	18.503049
2	-7.2858363	-17.829839
3	5.9622312	-44.786498
4	-2.9367652	-79.600335
5	7.0157385	100.708530
6	-7.0176886	-11.607554
7	6.6488317	-76.685980
8	2.0273459	-4.184077
9	-7.5505414	-10.278813
10	-11.049462	158.350190
11	-5.5037426	137.909280
12	-4.0652097	-65.033543
13	-6.3311991	-22.997814
14	2.7628582	3.620668
15	-8.1009221	-14.377606
16	5.7144264	71.165760
17	20.0824860	-28.422228
18	2.2988909	-84.930362
19	-5.8267679	24.920744
20	2.9184153	-54.443579

These residuals can then be plotted against the order in which the observations were taken. The plots are given in the guided solution.

There is no clear pattern in either plot that would suggest the subject got better in later trials due to learning or got worse due to fatigue. Time order does not appear to be a lurking variable.

Exercise 2.67

The value of $r = 0.999$.

This correlation is based on averaged data, namely the average stride rates of 21 runners at each of the seven values of speed (note that such data might have been collected by having the runners run on a treadmill, where speed can be controlled). If we had the data on the individual stride rates of all 21 runners, we would expect the correlation to decrease (be less than 0.999).

SECTION 2.5

OVERVIEW

A variable displays **linear growth** when it increases by the addition of a fixed positive amount in each equal time period. A scatterplot of such a variable versus time will look approximately linear. The methods of least-squares regression can be used to explore linear growth of a variable over time. Plots of residuals can be used to detect deviations in the overall linear pattern.

A variable displays **exponential growth** when it increases by multiplication by a fixed amount greater than 1 in each equal time period. A scatterplot of the **logarithm** of such a variable will look approximately linear. Exponential growth over time can thus be explored by applying the methods of least-squares regression to the logarithm of the variable. Plots of residuals can be used to detect deviations in the overall linear pattern of the logarithm, suggesting deviations in the overall pattern of exponential growth.

Linear and exponential growth are the two most common patterns of growth in the values of a variable over time.

GUIDED SOLUTIONS

Exercise 2.73

KEY CONCEPTS - exact exponential growth

a) A return of 11.34% means that if you invest X dollars, you will have 1.1134X dollars next year, $(1.1134)^2$X two years later, $(1.1134)^3$X dollars three years later, etc. Based on this pattern, how much will you have 25 years later (at the end of 1995)?

b) By reasoning similar to that in part (a), $1000 in 1970 is equivalent to how much 25 years later if the rate of inflation is 5.62%?

Exercise 2.75

KEY CONCEPTS - investigating exponential growth by looking at logarithms of data

a) Make a scatterplot of the data in the space provided below.

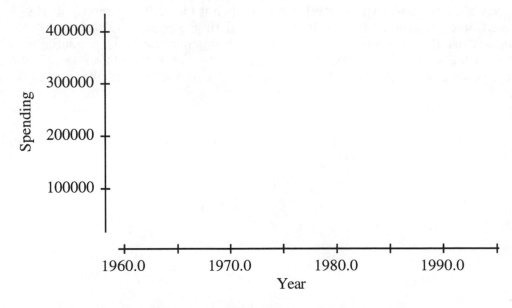

Is the pattern in the plot increasing in a linear or nonlinear manner?

b) While this can be done by hand, it is far easier to use statistical software. First take logarithms (base 10) of spending and use these in place of the original values of spending in your scatterplot and regression. Sketch your plot in the space below and draw in the regression line on the plot.

Year	Log(Spending)
1960	
1965	
1970	
1975	
1980	
1985	
1990	

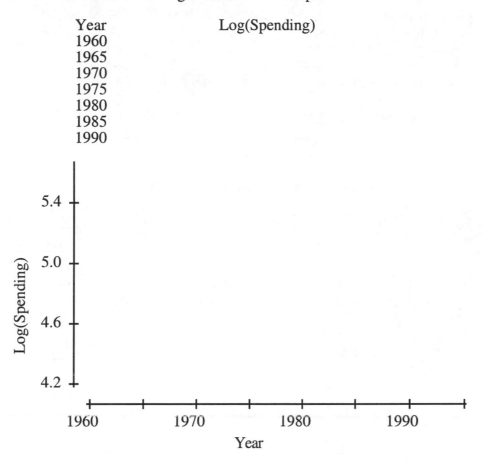

What features of your plot indicate periods of faster growth? Slower growth? Remember, the rate of growth is determined by the slope of the trend in the plot.

c) Use the equation of the least-squares regression line to predict the logarithm of spending. Be sure to convert this logarithm to a dollar amount before comparing it to the actual spending in 1992.

COMPLETE SOLUTIONS

Exercise 2.73

a) In general, if you invest X dollars, you will have $(1.1134)^n$X dollars n years later. Thus, if we invested $1000 in 1970, by 1995 we are $n = 25$ years later, so our investment would be worth

$$(1.1134)^{25}(\$1000) = (14.66494)(\$1000) = \$14,664.94$$

b) By the same reasoning as in (a), an inflation rate of 5.62% means that if we start with X dollars, after n years these X dollars are equivalent to $(1.0562)^n$X dollars. Thus, $1000 in 1970 is equivalent to

$$(1.0562)^{25}(\$1000) = (3.92332)(\$1000) = \$3,923.32$$

$n = 25$ years later in 1995.

Exercise 2.75

a) A scatterplot of the data with year as the explanatory variable and spending as the response is given below and is probably the best way to examine the data graphically.

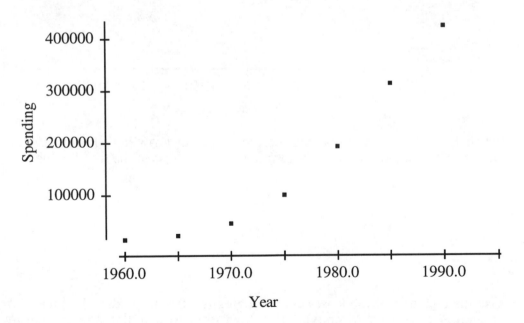

The pattern shows a curved, increasing trend and appears closer to exponential growth than to linear growth.

b) The logarithms of spending are given below.

Year	$\log_{10}$(spending)
1960	4.1555486
1965	4.3385959
1970	4.6555802
1975	4.9987605
1980	5.2814016
1985	5.4956942
1990	5.6255769

A scatterplot with year as the explanatory variable and log(spending) as the response is

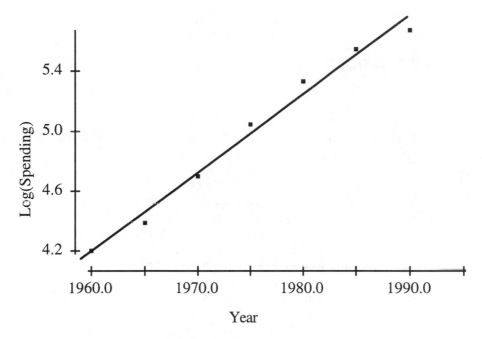

Using statistical software, the least-squares regression line in this case is

$$\log(\text{spending}) = -98.7531 + 0.0525(\text{year})$$

and this is plotted on the scatterplot above.

In this plot, faster growth than the overall trend is characterized by periods where the points seem to follow a line with a slope greater than the least-squares regression line, and slower growth by periods where the points seem to follow a line with a slope less than the least-squares regression line. From this plot, we see that

• Growth was faster than the overall trend from about 1965 to about 1980.

- Growth was slower than the overall trend from 1960 to 1965 and from about 1980 to 1990.

c) From the equation of our least-squares regression line, we would predict log(spending) for 1992 to be

$$\log(\text{spending}) = -98.7531 + 0.0525(1992) = 5.8269$$

which yields

$$\text{spending} = 10^{5.8269} = \$671,274.27$$

The amount actually spent in 1992 ($495,710) is less than the amount predicted by our least-squares regression line.

SECTION 2.6

OVERVIEW

This section discusses techniques for describing the relationship between two or more categorical variables. To analyze categorical variables, we use counts (frequencies) or percents (relative frequencies) of individuals that fall into various categories. A **two-way table** of such counts is used to organize data about two categorical variables. Values of the **row variable** label the rows that run across the table, and values of the **column variable** label the columns that run down the table. In each cell (intersection of a row and column) of the table, we enter the number of cases for which the row and column variables have the values (categories) corresponding to that cell.

The **row totals** and **column totals** in a **two-way table** give the marginal distributions of the two variables separately. It is usually clearest to present these distributions as percentages of the table total. Marginal distributions do not give any information about the relationship between the variables. **Bar graphs** are a useful way of presenting these marginal distributions.

The conditional distributions in a two-way table help us to see relationships between two categorical variables. To find the conditional distribution of the row variable for a specific value of the column variable, look only at that one column in the table. Express each entry in the column as a percentage of the column total. There is a conditional distribution of the row variable for each column in the table. Comparing these conditional distributions is one way to describe the association between the row and column variables, particularly if the column variable is the explanatory variable. When the row variable is explanatory, find the conditional distribution of the column variable for each row and compare these distributions. Side-by-side bar graphs of the conditional distributions of the row or column variable can be used to compare these distributions and describe any association that may be present.

Data on three categorical variables can be presented in a **three-way table**, printed as separate two-way tables for each value of the third variable. An association between two variables that holds for each level of this third variable can be changed, or even reversed, when the data are **aggregated** by summing over all values of the third variable. **Simpson's paradox** refers to a reversal of an association by aggregation.

GUIDED SOLUTIONS

Exercise 2.81

KEY CONCEPTS - marginal and conditional distributions in two-way tables

For convenience, we reproduce Table 2.15 below. The entries are in thousands of students)

Age	2-year full-time	2-year part-time	4-year full-time	4-year part-time
Under 18	36	98	75	37
18 to 21	1126	711	3270	223
22 to 34	634	1575	2267	1380
35 and up	221	1092	390	915
Total	2017	3476	6002	2555

a) Ask yourself: What numbers must be added from the entries in this table to determine how many undergraduate students were enrolled in colleges and universities?

b) Ask yourself: What ratio must you compute to determine the percentage of all undergraduate students that were 18 to 21 years old? How would you compute the numerator and denominator of this ratio from the information in the table? Both the numerator and denominator are sums of entries in the table. Remember to convert your ratio to a percentage.

numerator   =
denominator   =
percentage   =

c) Here, you must compute four ratios and then convert them to percentage. What totals go in the denominator? What numbers in the table go in the numerators of these ratios? To construct a bar graph representing these percentages, you may wish to use statistical software. If you wish to create the bar graph by hand, ask yourself what scale should be used for the vertical axis of the graph.

d) In which of the four categories are 18 to 21-year-olds the largest percentage? Do you see any pattern? Can you account for any patterns you observe?

Exercise 2.85

KEY CONCEPTS - conditional distributions and association in a two-way table

For each type of program (column of the table) you will have to convert entries to proportions of the column totals. Do this and fill in the numbers in the table below. Then ask yourself if there are any patterns in these associations. Can you give an explanation for any patterns you observe?

Age	2-year full-time	2-year part-time	4-year full-time	4-year part-time
Under 18				
18 to 21				
22 to 34				
35 and up				

Exercise 2.93

KEY CONCEPTS - aggregating three-way tables, Simpson's paradox

a) Create your two-way table by summing corresponding pairs of entries in the Business table and the Law table. Rows should indicate gender and columns admission status.

b) In your aggregated two-way table, convert row entries to percentages of the row totals. You should observe that a higher percentage of the male applicants than of the female applicants is admitted.

c) For each of the two original tables (the table for Business and the table for Law), convert row entries to percentages of the row totals. In each table, you should observe that a higher percentage of the female applicants than of the male applicants is admitted. This appears to contradict the results of (b)!

d) To explain the apparent contradiction observed in (c), consider which professional school is easier to get into and which professional school males and females tend to apply to. Phrase your answer in plain, clear English.

COMPLETE SOLUTIONS

Exercise 2.81

For convenience, we reproduce Table 2.15 below.

Age	2-year full-time	2-year part-time	4-year full-time	4-year part-time
Under 18	36	98	75	37
18 to 21	1126	711	3270	223
22 to 34	634	1575	2267	1380
35 and up	221	1092	390	915
Total	2017	3476	6002	2555

a) The total number of undergraduate students enrolled in colleges and universities in fall 1993 can be computed from the row labeled Total. This row provides the total number of students (in thousands of students) in each of the categories, so the sum of all the entries in this row (multiplied by 1000) gives the desired total. This sum is

total number of undergraduates
= 2,017,000 + 3,476,000 + 6,002,000 + 2,555,000
= 14,050,000

b) The desired percentage is

[(total number of 18 to 21 year olds) / (total number of undergraduate students)] × 100%

The total number of 18 to 21-year-olds is the sum of all the entries in the row labeled 18 to 21(multiplied by 1000). This total is

total number of 18 to 21 year olds
= 1,126,000 + 711,000 + 3,270,000 + 223,000
= 5,330,000.

In (a) we found

total number of undergraduates = 14,050,000

Thus, the percentage of all undergraduate students who were 18 to 21 years old in the fall of 1993 is

(5,330,000/14,050,000) × 100% = 37.936%

c) To compute these percentages, we need to determine what percentage each entry in the row labeled 18 to 21 is of the corresponding column total. These percentages are

2-year full time	[1126/2017] × 100% = 55.83%
2-year part time	[711/3476] × 100% = 20.45%
4-year full time	[3270/6002] × 100% = 54.48%
4-year part time	[223/2555] × 100% = 8.73%

A bar graph of these results is given below.

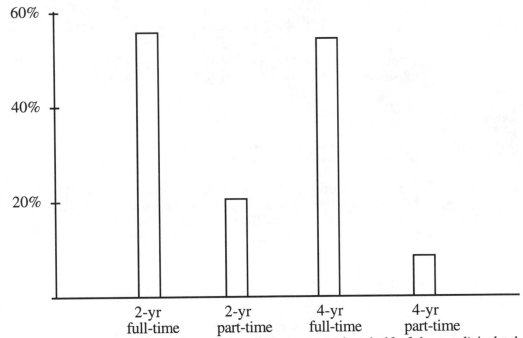

d) The 18-to-21 group predominates (makes up more than half of the total) in both of the full-time student categories. They are a much smaller percentage of the part-time student categories. These categories are dominated by older students who, presumably, work at least part-time while going to school.

Exercise 2.85

The conditional distributions of age for each type of program require that we compute the percentage that each entry in the table is of its column total. For example, to compute the conditional distribution of age for 2-year full-time students, we divide each entry in the column labeled 2-year full-time students by the column total of 2017 and convert the result to a percentage. If this is done for each column, we get the following.

Age	2-year full-time	2-year part-time	4-year full-time	4-year part-time
Under 18	1.78%	2.82%	1.25%	1.45%
18 to 21	55.83%	20.45%	54.48%	8.73%
22 to 34	31.43%	45.31%	37.77%	54.01%
35 and up	10.96%	31.42%	6.50%	35.81%

From these results, we see that the largest proportion of students in full-time programs tend to be from 18 to 21—those adults that we typically think of as being "college age." Presumably, these are younger people that can afford to be full-time students (have financial resources such as parental support so that they do not have to work, do not have a family to support, or do not have a job that they do not want to lose). Part-time programs tend to attract 22 to 34 and 35 or older students—those adults that we typically think of as being beyond college age. Presumably, these are somewhat older individuals who need to work to get through school, need to work to help support a family while in school, or have a job that they wish to retain while being in school.

Exercise 2.93

a) To make the desired table, we add together the entries in the corresponding cells of the two tables given. We get

	Admit	Deny
Male	480 + 10 = 490	120 + 90 = 210
Female	180 + 100 = 280	20 + 200 = 220

b) To compute the percentages, we must determine the row totals and then express each entry as a percentage of the row total. The row totals are 700 for the Admit column and 500 for the Deny column. We then obtain the following.

	Admit	Deny
Male	$(490/700) \times 100\% = 70\%$	$(210/700) \times 100\% = 30\%$
Female	$(280/500) \times 100\% = 56\%$	$(220/500) \times 100\% = 44\%$

We see that indeed a higher percentage of males is admitted than of females.

c) We repeat the type of calculations in (b) but applied separately to the original tables given. Note that the row totals for Business are 600 males and 200 females and for Law are 100 males and 300 females.

	Business	
	Admit	Deny
Male	$(480/600) \times 100\% = 80\%$	$(120/600) = 20\%$
Female	$(180/200) \times 100\% = 90\%$	$(20/200) = 10\%$

	Law	
	Admit	Deny
Male	$(10/100) \times 100\% - 10\%$	$(90/100) \times 100\% = 90\%$
Female	$(100/300) \times 100\% = 33.3\%$	$(200/300) \times 100\% = 66.7\%$

Now we see that each school admits a higher percentage of females than males.

d) If we examine the data more closely, we see that Business tends to admit a high percentage of students and Law tends to admit a low percentage of students. We also see that the majority of applicants to Business are male, whereas the majority of applicants to Law are female. The fact that more males than females apply to Business (where it is relatively easy to get admitted) and more females than males apply to Law (where it is relatively hard to get admitted) means that the overall admission rate for females will appear relatively low (since they are applying to Law, which is hard to get into) compared to the admission rate for males (since males are applying to Business, which is easy to get into).

SECTION 2.7

OVERVIEW

An observed association between two variables can be the result of several things. It can result from a **cause-and-effect** relationship between the variables. It can also be due to the effect of l**urking variables**; that is, variables not directly studied that may affect the response and possibly the explanatory variable. Lurking variables may operate through **common response**, in which case changes in both the explanatory and response variables are caused by changes in the lurking variable. Lurking variables may also cause **confounding**, in which case both the explanatory variable and the lurking variables cause changes in the response, but we cannot distinguish their individual effects.

The best way to determine if an association is the result of a cause-and-effect relationship between the explanatory variable and the response is through an **experiment** in which we control the influences of other variables. In the absence of good experimental evidence, be cautious in accepting claims of causation. Good evidence requires an association that appears consistently in many studies, a clear explanation for the alleged cause-and-effect relationship, and careful examination of possible lurking variables.

GUIDED SOLUTIONS

Exercise 2.99

KEY CONCEPTS - explaining causation, lurking variables

Ask yourself the following questions.

• Was the study an **experiment** in which the influences of other variables were controlled?

• If the study was not an experiment:

• Is there information that the observed association appears consistently in many studies?

• Is there a clear explanation for the alleged cause-and-effect   relationship?

• Is there evidence that possible lurking variables have been ruled out as possible causes?

Exercise 2.105

KEY CONCEPTS - lurking variables

Ask yourself what sorts of students are likely to study a foreign language for at least two years. How are such students likely to do in other subjects?

Exercise 2.107

KEY CONCEPTS - evidence for causation in studies that are not experiments

To determine what kinds of information you would seek in records, ask yourself:

• Would records contain any information to indicate that the observed association appears consistently in many studies or settings?

• Would the records provide a clear explanation for the alleged cause-and-effect relationship?

• Would information in the records allow you to identify or rule out possible lurking variables as causes?

COMPLETE SOLUTIONS

Exercise 2.99

The data are intriguing because they are based on such a large number of operations, but by themselves they do not prove that anesthetic C is causing more deaths than the others. The main weakness of the study is that it was not an experiment in which the influences of other factors were controlled. Some case might still be made for causation, but we would need to know that similar results have been observed in other studies, we would need a clear explanation of why anesthetic C might cause more deaths than the others, and we would need evidence that the effects of lurking variables have been ruled out. Unfortunately, no such information is given.

A possible lurking variable that needs to be ruled out is the types of operations for which the anesthetics are used. Is anesthetic C used more often than the others in operations involving long, difficult, and risky surgery? If so, this might explain why the death rate is higher for anesthetic C.

Exercise 2.105

The explanatory variable in this study is whether or not a student has studied a foreign language for at least two years. The response variable is scores on an English achievement test.

Probably the most important lurking variable is "quality of student" (how hardworking the student is, innate intelligence, innate language ability). Students who are willing to take at least two years of foreign language are also likely to be more serious or talented students who are likely to work hard or do well in other subjects.

Exercise 2.107

The records are not likely to provide us with a clear explanation of why anesthetic C would have a higher death rate. This probably requires a medical, biological, or chemical explanation.

The records would allow us to see if the anesthetics are used for different purposes. Perhaps certain anesthetics are used in simple, routine types of surgery, while others are used in complicated, more risky types of surgery. In general, we should look for other patterns of association to identify lurking variables.

The records would also allow us to see if the pattern is in a variety of settings, which would allow us to rule out possible lurking variables. We might look to see if the pattern repeats itself if we restrict to cases from specific geographic regions, specific types of hospitals, specific types of operations, or specific types of patients.

CHAPTER 3

PRODUCING DATA

SECTION 3.1

OVERVIEW

Chapters 1 and 2 describe methods for exploring data. Such **exploratory data analysis** is used to determine what the data tell us about the variables measured and their relations to one another. Conclusions apply to the data observed and may not generalize beyond these data.

Statistical inference produces answers to specific questions, along with a statement of how confident we are that the answer is correct. Answers are usually intended to apply beyond the data observed. This requires careful **production of data** appropriate for answering the specific questions asked.

Data can be produced in many ways. **Anecdotal data** based on a few isolated cases is usually unreliable. **Available data** collected for other purposes, such as data produced by government agencies, can be helpful but again is not always reliable. **Sampling** selects a part of a population of interest to represent the whole. Done properly, sampling can yield reliable information about a population. **Observational studies** are investigations in which one simply observes the state of some population, usually with data collected by sampling. Even with proper sampling, data from observational studies are generally not appropriate for investigating cause-and-effect relations between variables. **Experiments** are investigations in which data are generated by active imposition of some treatment on the subjects of the experiment. Properly designed experiments are the best way to investigate cause-and-effect relations between variables.

good distinction between inferential and descriptive

GUIDED SOLUTIONS

Exercise 3.1

KEY CONCEPTS - anecdotal data

Is this a pattern that has occurred regularly or an isolated event? Are there other reasons the stinkbugs might all be on the sunflowers, rather than preferring sunflowers to okra? In most gardens, sunflowers would be in another part of the garden from the okra. You might also look up information about stinkbugs.

Exercise 3.5

KEY CONCEPTS - explanatory and response variables, experiments

a) Think about what the medical team is trying to demonstrate with this study.

b) Remember, experiments are investigations in which data are generated by *active* imposition of some treatment on the subjects.

c) Think about what lurking variables might be present in this study. Consider the fact that mastectomies are an older form of treatment.

COMPLETE SOLUTIONS

Exercise 3.1

One isolated, anecdotal case is not a good basis for drawing general conclusions. There are several possible reasons that the stinkbugs might all be on the sunflowers. These include:

• The sunflowers were located in a part of the garden that was attractive to stinkbugs. If the okra were in this part of the garden, the stinkbugs would have been on the okra.

• Stinkbugs tend to collect in clusters, and it was pure chance that all were on the sunflowers. At another time, they might have all been on the okra.

• Stinkbugs eat garden plants such as okra but prefer certain types of insects. Perhaps there were such insects on the sunflower, which temporarily distracted the stinkbugs from eating the okra.

• Stinkbugs prefer sunflowers to okra.

Anecdotal evidence does not allow us to determine which (if any) of these might be the reason all the stinkbugs were on the sunflowers.

Exercise 3.5

a) The explanatory variables are the treatments (mastectomy versus removal of tumor and nearby lymph nodes followed by radiation). The response variable is the survival time after surgery.

b) This sounds like it might be an experiment because it involves examining the results of two treatments, but it is actually an observational study. The medical team simply examined the records of 25 large hospitals. In other words, they just observed a sample of available data. To be an experiment, the medical team itself would have had to impose the treatments on a collection of subjects and observe the subsequent survival times of these subjects.

c) This study will not show whether a mastectomy *causes* longer average survival time. The survival times for women on the two treatments may differ, but this difference may result from lurking variables such as

• Time of treatment. Mastectomies are an older treatment. Records on women having mastectomies are more likely to be older cases when recent effective drugs were not available and detection of the cancer did not occur as early.

• Size of tumor. Mastectomies may be more common in advanced cases in which the tumor is very large.

A properly designed experiment that controls for the effects of lurking variables is needed to determine if a mastectomy *causes* longer average survival time.

SECTION 3.2

OVERVIEW

Experiments are studies in which one or more **treatments** are imposed on experimental **units** or **subjects**. A treatment is a combination of levels of the explanatory variables, called **factors**. The **design** of an experiment is a specification of the treatments to be used and the manner in which units or subjects are assigned to these treatments. The basic features of well-designed experiments are **control**, **randomization**, and **replication**.

Control is used to avoid confounding (mixing up) the effects of treatments with other influences such as lurking variables. One such lurking variable is the **placebo effect**, which is the response of a subject to the fact of receiving any treatment. The simplest form of control is **comparative experimentation,** which involve comparisons between two or more treatments. One of these treatments may be a **placebo** (fake treatment), and those subjects receiving the placebo are referred to as a **control group**.

Randomization uses a well-defined chance mechanism to assign subjects to treatments. It is used to create treatment groups that are similar, except for chance variation, before application of treatments. Randomized, comparative experiments are used to prevent **bias**, or systematic favoritism of certain outcomes. **Tables of random digits** or computer programs that generate random numbers are well-defined chance mechanisms that are used to carry out randomization. In either case, numerical labels are assigned to experimental units and random numbers from the table or computer software determine which labels (units) are assigned to which treatments.

Replication, the use of many units in an experiment, reduces the effect of any chance variation between treatment groups arising from randomization. Replication increases the sensitivity of an experiment to differences in treatments.

Additional control in an experiment can be achieved by forming experimental units into **blocks** that are similar in some way that is thought to affect the response. In a **block design**, units are first formed into blocks and then randomization is carried out separately in each block. **Matched pairs** are a simple form of blocking used to compare two treatments. In a matched pairs experiment, either the same unit (the block) receives both treatments in a random order or very similar units are matched in pairs (the blocks). In the latter case, one member of the pair receives one of the treatments and the other member the remaining treatment. Members of a matched pair are assigned to treatments using randomization.

Good experiments require attention to details. **Double-blind** experiments are ones in which neither the subject nor the person measuring the response is aware of which treatment is being used. **Lack of realism** in an experiment can prevent us from generalizing the results.

GUIDED SOLUTIONS

Exercise 3.11

KEY CONCEPTS - identifying experimental units or subjects, factors, treatments, and response variables

You need to read the description of the study carefully. To identify the subjects, ask yourself exactly who was used in the study.

To identify the factors, ask yourself what question the experiment wished to answer. What variables does the description say the answer depends on? These are the factors.

To identify the treatments, what combinations of values of the factors were actually used in the experiment? These are the treatments.

To identify the response variables, ask yourself what was measured on the subjects after exposure to the treatments. These are the response variables. Note that there are several response variables in this experiment.

Exercise 3.15

KEY CONCEPTS - design of an experiment, randomization

a) To begin, identify the subjects, the factor, the treatments, and the response variable. Now outline your design. Be sure to specify

• How many treatments there are and hence how many groups of subjects you must form

• How you will assign subjects to treatment groups

• What the treatments are, that is, what each subject will be required to do

• What response you will measure and how you will decide if the treatments differ in their effect

You can outline your design in words or with a picture.

b) The list of names has been reproduced below. Assign a numerical label to each. Be sure to use the same number of digits for each label.

Abrams	Danielson	Gutierrez	Lippman	Rosen
Adamson	Durr	Howard	Martinez	Sugiwara
Afifi	Edwards	Hwang	McNeill	Thompson
Brown	Fluharty	Iselin	Morse	Travers
Cansico	Garcia	Janle	Ng	Turing
Chen	Gerson	Kaplan	Quinones	Ullman
Cortez	Green	Kim	Rivera	Williams
Curzakis	Gupta	Lattimore	Roberts	Wong

Now start reading line 121 in Table A. Read across the row in groups of digits equal to the number of digits you used for your labels (for example, if you used two digits for labels, read line 121 in pairs of digits). You will need to keep reading until you have selected all the names for the first treatment. This may require you to continue on to line 122, line 123, and subsequent lines. Since there are only two treatments in this experiment, you can stop after you have selected the names for treatment 1. The remaining names are assigned to treatment 2.

Exercise 3.25

KEY CONCEPTS - matched pairs design, randomization

The first thing you should do is identify the subjects, the factor, the treatments, and the response variable. Next, decide what are the matched pairs in this experiment. How will you use a coin flip to assign members of a pair to the treatments? What will you measure, and how will you decide whether the right hand tends to be stronger in right-handed people?

Exercise 3.27

KEY CONCEPTS - block designs

a) To assist you, we have arranged the subjects and their excess weight in order of increasing excess weight. Now decide which are the five blocks.

Williams 22	Santiago 27	Brunk 30	Jackson 33	Birnbaum 35
Festinger 24	Kendall 30	Obrach 30	Stall 33	Tran 35
Hernandez 25	Mann 28	Rodriguez 30	Brown 34	Nevesky 39
Moses 25	Smith 29	Loren 32	Dixon 34	Wilansky 42

b) Label the names in each block (you should need only the labels 1, 2, 3, 4 in each block since there are only four people in a block), choose a line in Table B, and start reading from left to right, assigning each member of a given block to one of the four treatments.

Write your results below, as indicated.

Line in Table B used =

Subjects on regimen A =

Subjects on regimen B =

Subjects on regimen C =

Subjects on regimen D =

Exercise 3.31

KEY CONCEPTS - properties of random digits

Write your answers (True or False) in the space provided. Remember that a table of random digits is defined to be a list of the digits 0, 1, 2, 3, 4, 5, 6, 7, 8, 9 that has the following properties

1. The digit in any position in the list has the same chance of being any one of 0, 1 2, 3, 4, 5, 6, 7, 8, 9.

2. The digits in different positions are independent in the sense that the value of one has no influence on the value of any other.

Additional properties are listed in the text below the box containing the definition above.

a) _____

b) _____

c) _____

COMPLETE SOLUTIONS

Exercise 3.11

The subjects are the undergraduate students who viewed the television program.

There are two factors in this study. One factor is the length of the ad (30 seconds or 90 seconds). The other factor is how often the ad is repeated (1, 3, or 5 times).

The treatments are the length - number of repetitions combinations. Thus there are six treatments (30 seconds and 1 repetition, 30 seconds and 3 repetitions, 30 seconds and 5 repetitions, 90 seconds and 1 repetition, 90 seconds and 3 repetitions, 90 seconds and 5 repetitions).

The response variables are the student answers to the questions about their recall of the ad, their attitude toward the camera, and their intention to purchase it. The number of response variables is equal to the number of questions.

Exercise 3.15

a) In this case, the subjects are the 40 unmarried women who are college seniors seeking employment. The factor is "information on child care," and there are two treatments. We label them as

Treatment 1: The pair of brochures for companies A and B, neither of which mentions child care.

Treatment 2: The pair of brochures for companies A and B in which the brochure for A says nothing about child care but the brochure for B describes the company's on-site child care facility.

The response variable is the company that a woman states she would prefer to work for.

The study should be done as follows. Women should be randomly assigned to treatments, with 20 being assigned to treatment 1 and the remainder to treatment 2. Each woman then reads the brochures and selects the company she prefers. One should probably make sure the order in which brochures are read is randomly determined. The proportion of women that choose company B for each treatment should be calculated and the results for the two groups compared. A picture that summarizes this procedure is given below.

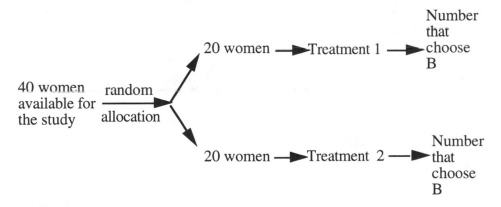

b) We label the 40 names using two-digit labels. We use the convention of starting with the label 00 and label down the columns. Of course, one could start with another number (such as 01) and label across rows if one wished. The names with labels are

00 Abrams	08 Danielson	16 Gutierrez	24 Lippman	32 Rosen
01 Adamson	09 Durr	17 Howard	25 Martinez	33 Sugiwara
02 Afifi	10 Edwards	18 Hwang	26 McNeill	34 Thompson
03 Brown	11 Fluharty	19 Iselin	27 Morse	35 Travers
04 Cansico	12 Garcia	20 Janle	28 Ng	36 Turing
05 Chen	13 Gerson	21 Kaplan	29 Quinones	37 Ullman
06 Cortez	14 Green	22 Kim	30 Rivera	38 Williams
07 Curzakis	15 Gupta	23 Lattimore	31 Roberts	39 Wong

Line 121 from Table B is reproduced below. We should read line 121 in pairs of digits from left to right. We have placed vertical bars between consecutive pairs to indicate how we have read the table. We underline those pairs that correspond to labels in our list and that have not been previously selected. We get

71|48|7 0|99|84| <u>29</u>|<u>07</u>|7 1|48|63| 6|68|<u>3 4</u>|70|52| 62|<u>22</u>|4 5|<u>10</u>|<u>25</u>

We find only six of our labels, so we need to continue reading on line 122:

<u>13</u>|87|<u>3 8</u>|<u>15</u>|98| 95|<u>05</u>|2 9|<u>09</u>|<u>08</u>| 73|59|<u>2 7</u>|51|86| 87|13|6 9|57|61

We still do not have enough labels, so we continue reading on line 123:

54|58|0 8|15|07| 27|10|2 5|60|27| 55|89|<u>2 3</u>|<u>30</u>|63| 41|84|<u>2 8</u>|18|68|

We still do not have enough labels, so we continue reading on line 124:

71|<u>03</u>|5 0|90|<u>01</u>| 43|<u>36</u>|7 494497 72719 96758 27611 91596

At 36, we reach 20 labels. We have reproduced our list of names with those selected underlined.

00 Abrams	<u>08 Danielson</u>	16 Gutierrez	24 Lippman	32 Rosen
<u>01 Adamson</u>	<u>09 Durr</u>	17 Howard	<u>25 Martinez</u>	33 Sugiwara
02 Afifi	<u>10 Edwards</u>	<u>18 Hwang</u>	26 McNeill	<u>34 Thompson</u>
<u>03 Brown</u>	11 Fluharty	19 Iselin	<u>27 Morse</u>	35 Travers
04 Cansico	12 Garcia	20 Janle	<u>28 Ng</u>	<u>36 Turing</u>
<u>05 Chen</u>	<u>13 Gerson</u>	21 Kaplan	<u>29 Quinones</u>	37 Ullman
06 Cortez	14 Green	<u>22 Kim</u>	<u>30 Rivera</u>	<u>38 Williams</u>
<u>07 Curzakis</u>	<u>15 Gupta</u>	<u>23 Lattimore</u>	31 Roberts	39 Wong

The 20 women whose names are underlined are assigned to treatment 1. The remaining 20 receive treatment 2.

Exercise 3.25

We have 15 subjects available. The factor is "hand" and there are two treatments. Treatment 1 is squeezing with the right hand and treatment 2 is squeezing with the left hand. The response is the force exerted as indicated by the reading on the scale.

To do the experiment, we use a matched pairs design. Each subject is a block, and each subject uses each hand (the matched pairs are the two hands of a particular subject). We should decide randomly which hand to use first, perhaps by flipping a coin. We measure the response for each hand and then compare the forces for the left and right hands over all subjects to see if there is a systematic difference between the two hands.

If we use a coin to do the randomization, we might decide that if we get heads the right hand goes first. Tails means the left hand goes first. For 15 flips of a coin we got

HTTHTTHTHHHTTTT

This tells us that subject 1 uses the right-hand first, subject 2 the left hand first, subject 3 the left hand first, and so forth. Notice with this scheme that the number of times the right hand goes first is six and the number of times the left hand goes first is nine.

Exercise 3.27

a) The subjects and their excess weights, rearranged in increasing order of excess weight, are listed below. The columns are the five blocks. We have labeled the subjects in each block from 1 to 4.

Block 1	Block 2	Block 3	Block 4	Block 5
1 Williams 22	1 Santiago 27	1 Brunk 30	1 Jackson 33	1 Birnbaum 35
2 Festinger 24	2 Kendall 30	2 Obrach 30	2 Stall 33	2 Tran 35
3 Hernandez 25	3 Mann 28	3 Rodriguez 30	3 Brown 34	3 Nevesky 39
4 Moses 25	4 Smith 29	4 Loren 32	4 Dixon 34	4 Wilansky 42

b) We used lines 130 and 131 in Table B, which is given below.

6905<u>1</u> 6<u>4</u>817 87174 09517 8<u>1</u>53<u>4</u> 06489 87<u>2</u>0<u>1</u>1 97<u>24</u>5

05007 <u>1</u>6632 8<u>11</u>94 14873<u>1</u> 04197 85576 45195 96565

For block 1, we read these from left to right, one digit at a time. The first label we encounter is assigned to regimen A, the next to regimen B, the next to regimen C, and the remaining label is then automatically assigned to regimen D. We underline those that are one of our labels, skipping repeats. The vertical lines indicates when we have completed a block. We summarize our results below.

Regimen A = Williams, Smith, Obrach, Brown, Birnbaum

Regimen B = Moses, Kendall, Loren, Stall, Wilansky

Regimen C = Hernandez, Santiago, Brunk, Jackson, Nevesky

Regimen D = Festinger, Mann, Rodriguez, Dixon, Tran

Exercise 3.31

a) This is false. Randomness does not mean each digit appears the exact same number of times in each row. For example, look at line 150 in Table B. There are only two 0s in this row.

b) True. Randomness means that each digit has an equal chance(1/10) of being a 0, each pair an equal chance(1/100) of being 00, each triple an equal chance(1/1000) of being 000, and so on. This point is made in the section "how to randomize" in the text a few paragraphs below the box containing the definition of a table of random digits.

c) False. Following the logic in (b), any set of four digits has an equal chance (1/10000) of being 0000. Thus, the digits 0000 can appear, but the chance of any four digits being this sequence is quite small.

SECTION 3.3

OVERVIEW

The **population** is the entire group of individuals or objects about which we want information. The information collected is contained in a **sample,** which is the part of the population we actually get to observe. How the sample is chosen (that is, the **design**) has a large impact on the usefulness of the data. A useful sample will be representative of the population and will help answer our questions. "Good" methods of collecting a sample include the following:

> **Probability samples**
> **Simple random samples,** also called **SRS**
> **Stratified random samples**
> **Multistage samples**

All these sampling methods involve some aspect of randomness through the use of a formal chance mechanism. Random selection is just one precaution that a person can take to reduce **bias,** the systematic favoring of a certain outcome. Samples we select using our own judgment, because they are convenient, or "without forethought" (mistaking this for randomness) are usually biased in some way. This is why we use computers or a tool like a **table of random digits** to help us select a sample.

A **voluntary response sample** includes people who choose to be in the sample by responding to a general appeal. They tend to be biased, because the sample is overrepresented by individuals with strong opinions, which are often negative.

Other kinds of bias to be on the lookout for include

nonresponse bias which occurs when individuals who are selected do not participate or cannot be contacted,

undercoverage which occurs when some group in the population is given either nochance or a much smaller chance than other groups to be in the sample, and

response bias which occurs when individuals do participate but are not responding truthfully or accurately due to the way the question is worded, the presence of an observer, fear of a negative reaction from the interviewer, or any other such source.

These types of bias can occur even in a randomly chosen sample and we need to try to reduce their impact as much as possible.

GUIDED SOLUTIONS

Exercise 3.33

KEY CONCEPTS - populations and nonresponse

Try and identify the population as exactly as possible by trying to identify exactly which individuals fall in the population. Where the information is not complete, you may need to make assumptions to try to describe the population in a reasonable way. Make sure not to confuse the population of interest with the population actually sampled. When they don't coincide there is always a strong potential for bias.

Exercise 3.37

KEY CONCEPTS - selecting a SRS with a table of random numbers

The table of random numbers can be used to select a SRS of numbers - in order to use it to sample from the students in the statistics course, the individuals in the course need to be assigned numbers. So that everyone does the problem the "same" way, we have first numbered the students according to alphabetical order in the list.

01- Agarwal	08 - Dewald	15 - Hixon	22 - Puri
02 - Anderson	09 - Fernandez	16 - Klassen	23 - Rodriguez
03 - Baxter	10 - Frank	17 - Liang	24 - Rubin
04 - Bowman	11 - Fuhrman	18 - Moser	25 - Santiago
05 - Bruvold	12 - Goel	19 - Naber	26 - Shen
06 - Casella	13 - Gupta	20 - Petrucelli	27 - Shyr
07 - Choi	14 - Hicks	21 - Pliego	28 - Sundheim

If you go to line 139 in the table and start selecting two digit numbers, then you should get the same answer as given in the complete solution.

Exercise 3.41

KEY CONCEPTS - systematic sampling

a) This is like the example except that there are now 200 addresses instead of 100, and the sample size is now five instead of four. With these two changes, you need to think about how many different systematic samples there are. Two different systematic samples are

systematic sample 1 =   01, 41, 81, 121, 161
systematic sample 2 =   02, 42, 82, 122, 162

....

How many systematic samples are there altogether? Choosing one of these systematic samples at random is equivalent to choosing the first address in the sample. The remaining four addresses follow automatically by adding 40. Carry out this procedure using line 120 in the table.

b) Why are all addresses equally likely to be selected? First, how many systematic samples contain each address? The chance of selecting an address is the same as the chance of selecting the systematic sample that contains it. With this in mind, what is the chance of any address being chosen? By the definition of an SRS, all samples of five addresses are equally likely to be selected. In a systematic sample, are all samples of five addresses even possible?

Exercise 3.45

KEY CONCEPTS - sampling frame, undercoverage

a) Which households wouldn't be in the sampling frame? Make some educated guesses about how these households might differ from those in the sampling frame (other than the fact that they don't have a phone number in the directory).

b) Random digit dialing makes the sampling frame larger—which households are added to it?

Exercise 3.49

KEY CONCEPTS - wording of questions

Questions can be worded to make it seem as though any reasonable person should agree (disagree) with the statement. Even though both (A) and (B) address the same issue, the tone of the questions is different enough to elicit different responses. Try to decide which question should have a higher proportion favoring banning contributions.

COMPLETE SOLUTIONS

Exercise 3.33

The population in the study includes all employed adult women. (The exact geographical area is not specified, but you might guess it is the United States. It is also unclear what is meant by employed. Does this refer only to full-time employment, or are adult women working part-time to be included in the population as well?) It is important to be very clear about the population of interest before beginning a study.

The sample is obtained from a subset of this population. It is selected from the 520 members of the local business and professional women's club. These are the only people who could be selected for our sample. Any systematic differences in the opinions of the women in the population of interest and the women in the local business and professional women's club will bias the results.

The nonresponse refers to the portion of the sample that did not respond. In this case, 52 of the 100 women sampled did not return the questionnaire, so the nonresponse rate is 52%.

Exercise 3.37

To choose an SRS of six students to be interviewed, first label the members of the population by associating a two-digit number with each.

01 Agarwal	08 Dewald	15 Hixon	22 Puri
02 Anderson	09 Fernandez	16 Klassen	23 Rodriguez
03 Baxter	10 Frank	17 Liang	24 Rubin
04 Bowman	11 Fuhrman	18 Moser	25 Santiago
05 Bruvold	12 Goel	19 Naber	26 Shen
06 Casella	13 Gupta	20 Petrucelli	27 Shyr
07 Choi	14 Hicks	21 Pliego	28 Sundheim

Now enter Table B and read two-digit groups until six students are chosen. Starting at line 139:

55588 994<u>04</u> 70708 4<u>10</u>98 43563 56934 48394 5<u>1719</u>
<u>12</u>975 13258 <u>13</u>048

The selected sample is 04 - Bowman, 10 - Frank, 17 - Liang, 19 - Naber,   12 - Goel, and 13 - Gupta.

Exercise 3.41

a) We want to select 5 addresses out of 200, so we think of the 200 addresses as five lists of 40 addresses. We choose one address from the first 40, and then every 40th address after that. The first step is to go to Table B, line 120, and choose the first two-digit random number you encounter that is one of the numbers 01,..., 40.

<u>35</u>476

The selected number is 35, so the sample includes addresses numbered 35, 75, 115, 155, and 195.

b) Each individual is in exactly one systematic sample, and the systematic samples are equally likely to be chosen. In our previous example, there were 40 systematic samples, each containing 5 addresses. The chance of selecting any address is the chance of picking the systematic sample it is in, which is 1 in 40.

A simple random sample of size n would allow every set of n individuals an equal chance of being selected. Thus, in this exercise, using an SRS, the sample consisting of the addresses numbered 1, 2, 3, 4, and 5 would have the same probability of being selected as any other set of 5 addresses. For a systematically selected sample, all samples of size n do not have the same probability of being selected. In our exercise, the sample consisting of the addresses numbered 1, 2, 3, 4, and 5 would have zero chance of being selected, since the numbers of the addresses do not all differ by 40. The sample we selected in (a)—35, 75, 115, 155, and 195—had a 1 in 40 chance of being selected, so all samples of 5 addresses are not equally likely.

Exercise 3.45

a) Households omitted from the frame are those that do not have a telephone number listed in the telephone directory. The types of people who might be underrepresented are poorer (including homeless) people who cannot afford to have a phone, and the group of people who have unlisted numbers. It is harder to characterize this second group. As a group, they would tend to have more money because you need to pay to have your phone number unlisted, or the group might include more single women who do not want their phone numbers available and possibly people whose jobs put them in contact with large groups of people might harass them if their phone number was easily accessible.

b) People with unlisted numbers will be included in the sampling frame. The sampling frame would now include any household with a phone. One interesting point is that all households will not have the same probability of getting in the sample, as some households have multiple phone lines and will be more likely to get in the sample. So, strictly speaking, random digit dialing will not actually provide an SRS of households with phones—just an SRS of phone numbers!

Exercise 3.49

As with many pairs of questions, changes in wording of the "same" question can elicit different responses.

(A)   The wording implies that contributions from special interest groups are bad: huge sums of money will exchange hands for the benefit of the special interest groups.

(B)   The wording suggests that we would be undemocratic to deny special interest groups the same right to contribute money as private individuals and other groups.

We would guess that 80% favored banning contributions when presented with question (A) and 40% favored banning contributions when presented with question (B).

SECTION 3.4

OVERVIEW

Statistical inference is the technique that allows us to use the information in a sample to draw conclusions about the population. To understand the idea of statistical inference, it is important to understand the distinction between **parameters** and **statistics.** A **statistic** is a number we calculate based on a sample from the population: its value can be computed once we have taken the sample, but its value varies from sample to sample. A statistic is generally used to estimate a population **parameter,** which is a fixed but unknown number that describes the population.

Statistic vs. Parameter

The variation in a statistic from sample to sample is called **sampling variability**. It can be described through the **sampling distribution** of the statistic, which is the distribution of values taken by the statistic in all possible samples of the same size from the population. The sampling distribution can be described in the same way as the distributions we encountered in Chapter 1. Three important features are

- A measure of center
- A measure of spread
- A description of the shape of the distribution

The properties and usefulness of a statistic can be determined by considering its sampling distribution. If the sampling distribution of a statistic is centered (has its mean) at the value of the population parameter, then the statistic is **unbiased** for this parameter. This means that the statistic tends to neither overestimate nor underestimate the parameter.

Another important feature of the sampling distribution is its spread. If the statistic is unbiased and the sampling distribution has little spread or variability, then the statistic will tend to be close to the parameter it is estimating for most samples. The variability of a statistic is related to both the sampling design and the sample size n. Larger sample sizes give smaller spread (better estimates) for any sampling design. An important feature of the spread is that as long as the population is much larger than the sample (at least 10 times), the spread of the sampling distribution will depend primarily on the sample size, not the population size.

If the parameter **p** is the proportion of the population with a particular characteristic, then the statistic $\hat{\mathbf{p}}$, the proportion in the sample with this characteristic, is an unbiased estimator. Provided the samples are selected at random, **probability** theory can be used to tell us about the distribution of a statistic.

GUIDED SOLUTIONS

Exercise 3.53

KEY CONCEPTS - statistics and parameters

In deciding whether a number represents a parameter or a statistic, you need to think about whether it is a numerical characteristic of the population of interest or a numerical characteristic of the particular sample that was selected. Statistics vary from sample to sample; parameters are fixed numerical characteristics of the population.

Exercise 3.57

KEY CONCEPTS - variability of the sample proportion

a) "As long as the population is much larger than the sample (say, at least 10 times larger), the spread of the sampling distribution for a sample of fixed size n is approximately the same for any population size." You need to think about how this rule applies to this example.

b) Is the rule given in part (a) applicable here? Read it carefully.

Exercise 3.61

KEY CONCEPTS - sampling distributions

a) The table of random numbers contains the 10 digits, 0, 1, 2,..., 9, which are "equally likely" to occur in any position selected at random from the table. If we want an egg mass to be present 20% of the time, then two digits correspond to the presence of an egg mass and the remaining eight digits correspond to the absence of an egg mass. Does it matter which two digits correspond to the presence of an egg mass?

While any two digits could be used, so that everyone does the same thing, let's let the occurrence of the digits 0 or 1 correspond to the presence of an egg mass and the remaining digits correspond to the absence. Also, let's start on line 128 of the table:

<u>1</u>5689 <u>1</u>4227

These 10 random digits correspond to our 10 sample areas. There are two sample areas with egg masses (correspond to a digit of 0 or 1), so that $\hat{p} = 0.2$ for this sample.

b) For this part of the problem, everyone will be taking their 20 samples from different parts of the random number table. Some of you may know how to get random samples from your computer software. Work with your own samples here. Your answers will not agree exactly with that given in the complete solution, but the general pattern should be similar. If everyone took 2000 samples instead of 20, would the sampling distributions from person to person show more or less agreement?

COMPLETE SOLUTIONS

Exercise 3.53

43 unlisted numbers is a property of the sample of 100 numbers dialed. It is the value of a statistic. **52%** of all Los Angeles phones unlisted is a property of the population of phones and is the value of a parameter.

Exercise 3.57

a) The population is at least 10 times the sample size $n = 2000$ for each of the states. So the variability in the sample proportion based on $n = 2000$ will be approximately the same for the population size of any state.

b) The problem switches here. The rule applies to a given sample size: the variability of the sample proportion based on a fixed number of observations is approximately the same for any population size. Now the sample size will vary from state to state. For Wyoming, 10% of the population is a sample size of about $n = 4850$, and 10% of the population of California is a sample size of about $n = 320,000$. Since larger sample sizes give smaller spread, there will be differences in the variability of the sample proportion from state to state. California's sample proportion will be much less variable than Wyoming's.

Exercise 3.61

a) This was done in the guided solution.

b) These are the values of $\hat{p}$ in the 20 samples we obtained using the computer to generate random digits.

sample	$\hat{p}$
1	0.1
2	0.1
3	0.0
4	0.1
5	0.0
6	0.4
7	0.0
8	0.3
9	0.2
10	0.0
11	0.2
12	0.0
13	0.4
14	0.2
15	0.4
16	0.2
17	0.2
18	0.1
19	0.3
20	0.2

```
0.0 | 00000
0.1 | 0000
0.2 | 000000
0.3 | 00
0.4 | 000
```

The mean of the distribution is 0.17. The shape looks fairly symmetric with a center near 0.2. Your stem and leaf plot may look quite different from this—with only 20 samples, the distributions may vary quite a bit from person to person. If everyone took 2000 samples, which would require the sampling to be done using a computer, then the shapes of the distributions would be quite similar from person to person.

CHAPTER 4

PROBABILITY: THE STUDY OF RANDOMNESS

SECTION 4.1

OVERVIEW

A process or phenomenon is called **random** if its outcome is uncertain. Although individual outcomes are uncertain, when the process is repeated a large number of times, the underlying distribution for the possible outcomes begins to emerge. For any outcome, its **probability** is the relative frequency (proportion of times) with which the outcome would occur in a long series of repetitions of the process. It is important that these repetitions or trials be **independent** for this property to hold.

We can study random behavior by carrying out physical experiments such as coin tossing of rolling of a die, or we can simulate a random phenomenon on the computer. Using the computer is particularly helpful when we want to consider a large number of trials.

GUIDED SOLUTIONS

Exercise 4.7

KEY CONCEPTS – assigning probabilities

This is a good "game" to play to start learning about random behavior, and it is an exercise for which everyone will get different answers. What you need to do is list the 20 outcomes in terms of Betty's possible winnings (–$4, –$2, $0, $2, $4) and then complete the relative frequency table below. These relative frequencies represent guesses or estimates of the true probabilities of the different outcomes. Since the relative frequencies are based on only 20 trials, they will probably not be that close to the true probabilities, but you should be able to see which outcomes are more or less likely.

Outcome   Relative frequency
–$4
–$2
$0
$2
$4

Exercise 4.9

KEY CONCEPTS – simulating a random phenomenon

a) You will need to use your computer software to simulate the 100 trials. After you do this, the computer can be used to calculate the proportion of "hits."

Proportion of hits =

For most students, their proportion of hits will be within 0.05 or 0.10 of the true probability of 0.5.

b) You need to go through your sequence to determine the longest string of hits or misses.

Longest run of shots hit =   Longest run of shots missed =

COMPLETE SOLUTIONS

Exercise 4.7

Our 20 trials yielded the following outcomes.

0	0	0	2	0	0	–4	0	2	2
–2	2	2	–4	0	–2	0	2	2	0

The relative frequencies of the outcomes are given below. We have also computed the true probabilities (long–term relative frequencies) using methods that you will learn later in this chapter. The relative frequencies show the general pattern, but with only 20 trials the differences between the relative

frequencies and true probabilities are still quite large. The agreement would
improved if the number of trials were increased.

Outcome	Relative frequency	True probability
−4	2/20 = 0.10	0.0625
−2	2/20 = 0.10	0.2500
0	9/20 = 0.45	0.3750
2	7/20 = 0.35	0.2500
4	0/20 = 0.00	0.0625

Exercise   4.9

a) Our sequence of hits (H) and misses (M) is given below.

```
H   H   M   H   H   H   M   M   H   H   H   M   H   M   H
M   H   M   M   H   M   M   H   H   H   M   M   H   H   H
M   M   M   M   H   M   M   H   H   H   H   M   H   H   H
M   M   M   M   M   H   M   H   H   M   H   M   H   M   M
H   H   H   H   M   H   M   M   M   M   H   H   M   H   H
M   M   H   H   H   M   M   H   H   M   M   H   M   H   M
M   H   M   H   H   M   H   H   H   H
```

Proportion of hits = 0.54

b) You need to go through your sequence to determine the longest string of hits or
misses. In our example,

Longest run of shots hit = 4 (this occurred more than once)
Longest run of shots missed = 5

SECTION 4.2

OVERVIEW

The description of a random phenomenon begins with the **sample space,** which is
the list of all possible outcomes. A set of outcomes is called an **event.** Once we
have determined the sample space, a **probability model** tells us how to assign
probabilities to the various events that can occur. There are four basic rules that
probabilities must satisfy:
- Any probability is a number between 0 and 1.
- All possible outcomes together must have probability 1.
- The probability that an event does not occur is 1 minus the probability that
    the event occurs.
- If two events have no outcomes in common, the probability that one or the
    other occurs is the sum of their individual probabilities.

In a sample space with a finite number of outcomes, probabilities are assigned to the individual outcomes, and the probability of any event is the sum of the probabilities of the outcomes that it contains. In some special cases, the outcomes are all **equally likely** and the probability of any event A is just computed as

$$P(A) = \text{(number of outcomes in A/number of outcomes in S)}$$

Events are **disjoint** if they have no outcomes in common. In this special case the probability that one or the other event occurs is the sum of their individual probabilities. This is the addition rule for for disjoint events, namely,

$$P\ (A \text{ or } B) - P\ (A) + P\ (B)$$

Events are **independent** if knowledge that one event has occurred does not alter the probability that the second event will occur. The mathematical definition of independence leads to the **multiplication rule** for independent events. If A and B are independent, then

$$P\ (A \text{ and } B) = P\ (A)\ P\ (B)$$

In any particular problem we can use this definition to check if two events are independent by seeing if the probabilities multiply according to the definition. Most of the time, however, independence is assumed as part of the proabability model. The four basic rules, plus the multiplication rule, allow us to compute the probabilities of events in many random phenomena.

Many students confuse independent and disjoint events once they have seen both definitions. Remember, disjoint events have no outcomes in common and when two events are disjoint, you can compute $P\ (A \text{ or } B) = P\ (A) + P\ (B)$ in this special case. The probability being computed is that one *or* the other event occurs. Disjoint events cannot be independent, since once we know that A has occurred, then the probability of B occurring becomes 0 (B cannot have occurred as well—this is the meaning of disjoint). The multiplication rule can be used to compute the probability that two events occur *simultaneously*, $P(A \text{ and } B) = P\ (A)\ P\ (B)$, in the special case of independence.

[handwritten margin note: mutually exclusive and independent]

GUIDED SOLUTIONS

Exercise 4.13

KEY CONCEPTS – sample space

One of the main difficulties encountered when describing the sample space is finding some notation to express your ideas formally. Following the text, our general format is S = {   }, where a description of the outcomes in the sample space is included within the braces.

a) You want to express that any number between 0 and 24 is a possible outcome. So you would write S = {all numbers between 0 and 24}.

b) S =

c) S =

d) You may not want to put an upper bound on the amount, so you should allow for any number greater than 0. S =

e) Remember that the rats can lose weight. S =

Exercise   4.15

KEY CONCEPTS – applying the probability rules

a) Since these are the only blood types, what has to be true about the sum of the probabilities for the different types? Use this to find $P(AB)$.

b) What's true about the events O and B blood type? Which probability rule do we follow? (Don't be confused by the wording in the problem that says "people with blood types O *and* B." In the context of this problem and the language of probability we are using, it really means O or B. There are no people with blood types O and B.)

Exercise   4.25

KEY CONCEPTS – probability rules, two–way table of probabilities

a) To be a legitimate assignment of probabilities to a sample space, the probabilities of the individual outcomes need to be between 0 and 1 (0 and 1 included), and the probabilities of the individual outcomes need to add to 1.

b) Add up the probabilities of the outcomes making up the event that the worker is female.

c) It's easier to use the rule for complements.

d) Think about whether these two classes are disjoint.

e) Use the rule for complements again.

Exercise 4.27

KEY CONCEPTS – sample spaces for simple random sampling, probabilities of events

a) It's easy to make the list. S = {(Abby, Deborah), (Abby, Julie), etc.}. There is no need to include both (Abby, Deborah) and (Deborah, Abby) in your list, since both refer to the same two individuals.

b) How many outcomes are there in the sample space in (a)? If they are equally likely, what is the probability of each?

c) How many outcomes in S include Julie? When the outcomes are equally likely, the probability of the event is just the

(number of outcomes in the event)/(number of outcomes in S)

d) How many outcomes in S include neither of the two men?

Exercise 4.33

KEY CONCEPTS – multiplication rule for independent events

Look at the first five rolls of each of the sequences. They have the same probability, since there is one G and 4 R's. So for the first five rolls, none of the sequences would be preferred (they all have the same probability). On the sixth roll, in the first sequence you could get either a G or an R and you would win. For the second sequence you need a G on the sixth roll to win, and for the third sequence you need an R. So this makes the first sequence have the highest probability. Why do you think most students chose the second sequence? To check, compute the probabilities for each of the three sequences using the multiplication rule.

P(First sequence) =

P(Second sequence) =

P(Third sequence) =

Exercise 4.35

KEY CONCEPTS - multiplication rule for independent events

a) The three years are independent. If U indicates a year for the price being up and D indicates a year for the price being down, you need to compute P(UUU).

b) Since the events are independent, what happens in the first two years does not affect the probability of going up or down in the third year. What's the probability of the price going down in any given year?

c) This problem must be set up carefully and done in steps.

Step 1 - Write the event of interest in terms of simpler outcomes. How would you write P(UU or DD) in terms of P(UU) and P(DD)?

P(moves in the same direction in the next two years) = P(UU or DD).

Step 2 - Evaluate P(UU) and P(DD) and substitute your answer in the expression from step 1.

COMPLETE SOLUTIONS

Exercise 4.13

a) S = {all numbers between 0 and 24}.
b) S = {0, 1, 2, ..., 11,000}.
c) S = {0, 1, ..., 12}.
d) S = {set of all numbers greater than or equal to zero}.
c) S = {all positive and negative numbers}. The inclusion of the negative numbers allows for the possibility that the rats lose weight.

Exercise 4.15

a) The probabilities for the different blood types must add to 1. The sum of the probabilities for blood types O, A, and B is $0.49 + 0.27 + 0.20 = 0.96$. Subtracting this from 1 tells us that the probability of the remaining type AB must be 0.04.

b) Maria can receive transfusions from people with blood types O or B. Since a person cannot have both of these blood types, they are disjoint. The calculation follows probability rule 4, which says that $P(\text{O or B}) = P(\text{O}) + P(\text{B}) = 0.49 + 0.20 = 0.69$.

Exercise 4.25

a) All probabilites are between 0 and 1. The sum is

$$0.14 + 0.11 + 0.06 + 0.11 + 0.12 + 0.03 + 0.09 + 0.20 + 0.08 \\ + 0.01 + 0.04 + 0.01 = 1.00$$

b) The outcomes in the second row correspond to the proportion of workers who are female and in each of the occupation classes. Just add up these numbers to get

$$P(\text{female}) = 0.09 + 0.20 + 0.08 + 0.01 + 0.04 + 0.01 = 0.43$$

c) The probability of being in farming, forestry or fishing (occupation class F) is $P(\text{F}) = 0.03 + 0.01 = 0.04$ (add up the proportion of males and females in this occupation class). The problem asks for the complementary event (not in farming, forestry, or fishing), which follows from the probability rule $P(\text{F}^C) = 1 - P(\text{F}) = 1 - 0.04 = 0.96$.

d) The two events, occupational class D and occupational class E, are disjoint. So the probability of holding a job in once of these classes follows from the probability rule:

$$P(\text{D or E}) = P(\text{D}) + P(\text{E}) = 0.12 + 0.16 = 0.28$$

Note: $P(\text{D}) = 0.11 + 0.01 = 0.12$ and a similar calculation gives $P(\text{E})$.

e) This is the complement of the event in (d), so we subtract the answer in (d) from 1 to get $1 - 0.28 = 0.72$.

Exercise 4.27

a) S = {(Abby, Deborah), (Abby, Julie), (Abby, Sam), (Abby, Roberto), (Deborah, Julie), (Deborah, Sam), (Deborah, Roberto), (Julie, Sam), (Julie, Roberto), (Sam, Roberto)}.

b) There are 10 possible outcomes. Since they are equally likely, each has probability 0.10.

c) Julie is in four of the outcomes, so her chance of attending the conference in *Paris* is 4/10 = 0.4. (Note that each person has the same probability of going. Remember from Chapter 2 that this is a property of an SRS.)

d) The chosen group must contain two women, (Abby, Deborah), (Abby, Julie), or (Deborah, Julie). There are three possibilities, so the desired probability is 3/10 = 0.3.

Exercise 4.33

You need to use the multiplication rule because the rolls are independent. For the first sequence,

$$P(\text{RGRRR}) = \left(\frac{2}{6}\right)\left(\frac{4}{6}\right)\left(\frac{2}{6}\right)\left(\frac{2}{6}\right)\left(\frac{2}{6}\right).$$

In the product for the probability of this sequence, the 2/6 appears 4 times, once for each R outcome, and the 4/6 appears once, for the G outcome.

$$P(\text{First sequence}) = \left(\frac{2}{6}\right)^4\left(\frac{4}{6}\right) = 0.0082$$

$$P(\text{Second sequence}) = \left(\frac{2}{6}\right)^4\left(\frac{4}{6}\right)^2 = 0.0055$$

$$P(\text{Third sequence}) = \left(\frac{2}{6}\right)^5\left(\frac{4}{6}\right) = 0.0027$$

Exercise 4.35

a) $P(\text{UUU}) = (0.65)^3 = 0.2746$.

b) The probability of the price being down in any given year is $1 - 0.65 = 0.35$. Since the years are independent, the probability of the price being down in the third year is 0.35, regardless of what has happened in the first two years.

c) $P(\text{moves in the same direction in the next two years}) = P(\text{UU or DD}) = P(\text{UU}) + P(\text{DD})$, since the events UU and DD are disjoint. Using the independence of two successive years, $P(\text{UU}) = (0.65)^2 = 0.4225$, and $P(\text{DD}) = (0.35)^2 = 0.1225$. Putting this together,

$P(\text{moves in the same direction in the next two years}) = 0.4225 + 0.1225 = 0.5450$.

SECTION 4.3

OVERVIEW

A **random variable** is a variable whose value is a numerical outcome of a random phenomenon. The restriction to numerical outcomes makes the description of the probability model simpler and allows us to begin to look at some further properties of probability models in a unified way. If we toss a coin three times and record the sequence of heads and tails, then an example of an outcome would be HTH, which would not correspond directly to a random variable. On the other hand, if we were only keeping track of the number of heads on the three tosses, then the outcome of the experiment would be 0, 1, 2, or 3 and would correspond to the value of the random variable X = number of heads.

The two types of random variables we will encounter are **discrete** and **continuous** random variables. The **probability distribution** of a random variable tells us about the possible values of X and how to assign probabilities to these values. A discrete random variable has a finite number of values, and the probability distribution is a list of the possible values of X and the probabilities assigned to these values. The probability distribution can be given in a table or using a **probability histogram**. For any event described in terms of X, the probability of the event is just the sum of the probabilities of the values of X included in the event.

A continuous random variable takes all values in some interval of numbers. Probabilities of events are determined using a **density curve**. The probability of any any event is the area under the curve corresponding to the values that make up the event. For density curves that involve regular shapes such as rectangles or triangles, we can compute probabilities of events using simple geometrical arguments. The **normal distribution** is another example of a continuous probability distribution, and probabilties of events for normal random variables are computed by standardizing and referring to Table A, as was done in Section 1.3.

GUIDED SOLUTIONS

Exercise 4.39

KEY CONCEPTS – discrete random variables, computing probabilities

a) Write the event in terms of a probability about the random variable X. Although you can figure out the answer without doing this, it's good practice to start using the notation for random variables.

b) Check to see that all the probabilities are numbers between 0 and 1 and that they sum to 1.

c) Add up the probabilities for values of $X \leq 3$.

d) Add up the probabilities for values of $X < 3$. Make sure that you understand the difference between this answer and what was required for part (c).

e) Events about X are typically "that is bigger than or bigger than or equal to some number," or "that is less than or less than or equal to some number." Write the event in one of these forms.

Exercise 4.43

KEY CONCEPTS – finding the probability distribution of a random variable

a) The probability of a randomly selected student opposing the funding of interest groups is 0.4 and the probability of favoring it is 0.6. The opinions of different students sampled are independent of one another. So you can use the multiplication rule to find

P(A supports, B supports, and C opposes) =

b) It is easiest to do this by making a table to keep track of the calculations. The first entry is given below. There should be eight lines to the table when you're done. If you've done the calculations correctly, what should be true about the eight probabilities? Don't worry about the column labeled "Value of X" for now. It will not be needed until part (c).

A	B	C	Probability	Value of X
support	support	support	$(0.6)^3 = 0.216$	

c) For each committee listed in the table in (b), find the associated value of X. For the first row, 0 people oppose the funding of interest groups, so the value of

X is 0 for a committee with these views. The values of X that can occur are 0, 1, 2, and 3. To find the probability that X takes any of these values, just add up the probabilities of the committees with that value of X. Fill in the table below with your values and make sure that the probabilities sum to 1.

Value of X	Probability

d) If a majority opposes funding, how many people on the committee would have to oppose funding? What does this say about X? Now use the table you constructed in (c) to evaluate this probability.

Exercise 4.45

KEY CONCEPTS – continuous random variables, computing probabilities

a) As with finding areas under normal curves, it helps to draw a sketch of the density that includes the area corresponding to the probability that you need to evaluate. In this part you need to find $P(X \le 49)$ when X has the uniform density curve. The density curve with the area corresponding to this probability is given below. Since it is a rectangular region, the area corresponds to the length $\times$ height $= 0.49 \times 1 = 0.49$, which is the $P(X \le 0.49)$. Remember that for continuous densities, $P(X \le 0.49) = P(X < 0.49)$.

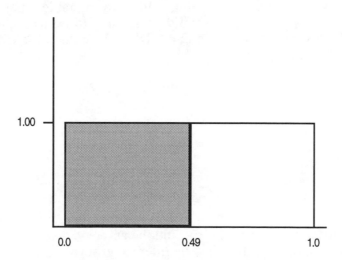

b) Sketch the density and the area you need below.

c) This interval includes values that have no probability since $X \leq 1$. So the values greater than 1 do not contribute to the probability.

d) Sketch the density and the area you need below.

e) Sketch the density and the area you need below.

f) What's true about the probability of an individual outcome for a continuous random variable?

Exercise 4.49

KEY CONCEPTS – probabilities for a sample proportion

a) Finding probabilities associated with a sample proportion when we know the mean and standard deviation of the sampling distribution requires first standardizing to a z–score so that we can refer to the table for the standard normal distribution. In this example, the mean is $\mu = 0.15$, (the value of p), and the standard deviation is given as $\sigma = 0.0092$. If you are still uncomfortable doing this type of problem, it is best to continue to draw a picture of a normal curve and the required area as we did in Section 1.3. Otherwise, you can just follow the method of Example 4.18 in this chapter.

$P(\hat{p} \geq 0.16) =$

b) $P(0.14 \leq \hat{p} \leq 0.16) =$

COMPLETE SOLUTIONS

Exercise 4.39

a) $P(X = 5) = 0.01$

b) All the probabilities are between 0 and 1, and the sum of the probabilities is $0.48 + 0.38 + 0.08 + 0.05 + 0.01 = 1.00$

c) For the probability that $X \leq 3$, add the probabilities corresponding to $X = 1, 2,$ and 3 to give $P(X \leq 3) = 0.48 + 0.38 + 0.08 = 0.94$.

d) The probability for $X < 3$ only includes values of X that are strictly below 3, which corresponds to $X = 1$ and 2. So, $P(X < 3) = 0.48 + 0.38 = 0.86$.

e) The son reaches one of the two higher classes if X is equal to 4 or 5. In terms of a single expression, this event is $X \geq 4$. Then $P(X \geq 4) = 0.05 + 0.01 = 0.06$. (Notice that this is the complement of the event in (c), which is why the two probabilities sum to 1.)

Exercise 4.43

a) P(A supports, B supports and C opposes)
     $= P$(A supports) P(B supports) P(C opposes) $= (0.6)(0.6)(0.4) = 0.144$.

b)

A	B	C	Probability		Value of X
support	support	support	$(0.6)^3$	$= 0.216$	0
support	support	oppose	$(0.6)^2(0.4)$	$= 0.144$	1
support	oppose	support	$(0.6)^2(0.4)$	$= 0.144$	1
oppose	support	support	$(0.6)^2(0.4)$	$= 0.144$	1
support	oppose	oppose	$(0.6)(0.4)^2$	$= 0.096$	2
oppose	support	oppose	$(0.6)(0.4)^2$	$= 0.096$	2
oppose	oppose	support	$(0.6)(0.4)^2$	$= 0.096$	2
oppose	oppose	oppose	$(0.4)^3$	$= 0.064$	3
			Total	$= 1.00$	

c) The value of X is given in the table in (b). The possible values of X are 0, 1, 2, and 3. To find the probability that X takes any value, just add up the probabilities of the committees with that value of X. For example, $P(X = 2) = 3(0.096) = 0.288$.

Value of X	Probability
0	0.216
1	0.432
2	0.288
3	0.064

d) A majority oppose if either two or three members of the board oppose. If two oppose, then $X = 2$, and if three oppose, then $X = 3$. So the event is $X \geq 2$ and the required probability is $P(X \geq 2) = 0.288 + 0.064 = 0.352$.

Exercise   4.45

a) See the guided solution.

b) The shaded area is $P(X \geq 0.27) = (0.83)(1) = 0.83$.

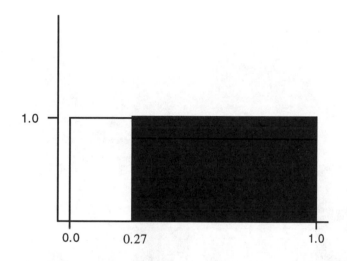

c) $P(0.27 < X < 1.27) = P(0.27 < X < 1.00) = P(X > 0.27)$. This is the probability computed in (b), since $P(X > 0.27) = P(X \geq 0.27)$.

d)

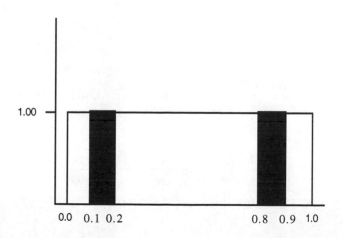

$P(0.1 \leq X \leq 0.2 \text{ or } 0.8 \leq X \leq 0.9) = P(0.1 \leq X \leq 0.2) + P(0.8 \leq X \leq 0.9)$, since the two intervals are disjoint.

$P(0.1 \leq X \leq 0.2) = (0.1)(1) = 0.1$ and $P(0.8 \leq X \leq 0.9) = (0.1)(1) = 0.1$.

Putting these together gives $P(0.1 \leq X \leq 0.2 \text{ or } 0.8 \leq X \leq 0.9) = 0.1 + 0.1 = 0.2$.

e) In the picture, the area shaded includes all values of X that are not in the interval 0.3 to 0.8.

$P(X$ is not in the interval 0.3 to 0.8$) = P(0 \leq X \leq 0.3) + P(0.8 \leq X \leq 1) = 0.3 + 0.2 = 0.5$.

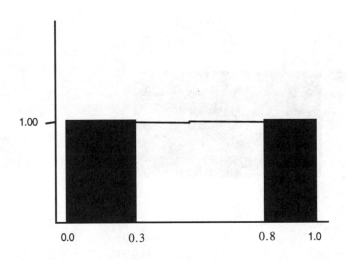

f) Since X has a continuous distribution, the probability of a single point is zero. So, $P(X = 0.5) = 0$.

Exercise 4.49

a)

$$P(\hat{p} \geq 0.16) = P\left(\frac{\hat{p} - 0.15}{0.0092} \geq \frac{0.16 - 0.15}{0.0092}\right) = P(Z \geq 1.09) = 1 - 0.8621 = 0.1379$$

b)

$$P(0.14 \leq \hat{p} \leq 0.16) = P\left(\frac{0.14 - 0.15}{0.0092} \leq \frac{\hat{p} - 0.15}{0.0092} \leq \frac{0.16 - 0.15}{0.0092}\right)$$
$$= P(-1.09 \leq Z \leq 1.09) = 0.8621 - 0.1379 = 0.7242$$

SECTION 4.4

OVERVIEW

In Chapter 1, we introduced the concept of the distribution of a set of numbers or data. The distribution describes the different values in the set and the frequency or relative frequency with which those values occur. The mean of the numbers is a measure of the center of the distribution, and the standard deviation is a measure of the variability or spread. These concepts are also used to describe features of a random variable X. The probability distribution of a random variable indicates the possible values of the random variable and the probability (relative frequency in repeated observations) with which they occur. The **mean** μ_X of a random variable X describes the center or balance point of the probability distribution or density curve of X. If X is a discrete random variable having possible values x_1, $x_2,\ldots,x_k$ with corresponding probabilites $p_1, p_2,\ldots,p_k$, the mean μ_X is the average of the possible values weighted by the corresponing probabilities: that is,

$$\mu_X = x_1 p_1 + x_2 p_2 + \ldots + x_k p_k$$

The mean of a continuous random variable is computed from the density curve, but computations require more advanced mathematics. The law of large numbers relates the mean of a set of data to the mean of a random variable and says that the average of the values of X observed in many trials approaches μ_X.

The **variance** σ_X^2 of a random variable X is the average squared deviation of the values of X from their mean. For a discrete random variable,

$$\sigma_X^2 = (x_1 - \mu_X)^2 p_1 + (x_2 - \mu_X)^2 p_2 + \ldots + (x_k - \mu_X)^2 p_k.$$

The **standard deviation** σ_X is the positve square root of the variance. The standard deviation measures the variability of the distribution of the random variable X about its mean. The variance of a continuous random variable, like the mean, is computed from the density curve. Again, computations require more advanced mathematics.

The mean and variances of random variables obey the following rules. If a and b are fixed numbers, then

$$\mu_{a + bX} = a + b\mu_X$$

$$\sigma^2_{a + bX} = b^2 \sigma^2_X$$

If X and Y are any two random variables, then

$$\mu_{X+Y} = \mu_X + \mu_Y$$

If X and Y are independent random variables, then

$$\sigma^2_{X+Y} = \sigma^2_X + \sigma^2_Y$$

$$\sigma^2_{X-Y} = \sigma^2_X + \sigma^2_Y$$

GUIDED SOLUTIONS

Exercise 4.51

KEY CONCEPTS – mean of a random variable

Recall that the average (mean) of X = grade in this course is computed using the formula

$$\mu_X = x_1 p_1 + x_2 p_2 + \ldots + x_k p_k$$

where the values of x_i and the p_i are given in the following table:

Grade (x)	0	1	2	3	4
Probability (p)	0.10	0.15	0.30	0.30	0.15

Now use the formula to compute μ_X.

Exercise 4.57

KEY CONCEPTS – independence, misconceptions about the law of large numbers

The key concept that must be properly understood to answer the questions raised in this problem is the notion of independence. Events A and B are independent if knowledge that A has occurred does not alter our assessment of the probability that B will occur. Do not be misled by misconceptions based on a faulty understanding of the nature of random behavior or a faulty understanding of the law of large numbers (either that runs indicate that a hot streak is in progress and will continue for a while, or that a run of one type of outcome must be immediately balanced by a lack of the outcome for several trials).

Exercise 4.61

KEY CONCEPTS – variance and standard deviation of a random variable

Recall that the variance of X = grade in this course is computed using the formula

$$\sigma_X^2 = (x_1 - \mu_X)^2 p_1 + (x_2 - \mu_X)^2 p_2 + \ldots + (x_k - \mu_X)^2 p_k$$

where μ_X is the mean of X (calculated in Exercise 4.51 as $\mu_X = 2.25$) and the values of x_i and the p_i are given in the following table:

Grade (x)	0	1	2	3	4
Probability (p)	0.10	0.15	0.30	0.30	0.15

Now use the formula to compute $\sigma^2{}_X$. To compute the standard deviation σ_X take the positive square root of the variance.

Exercise 4.63

KEY CONCEPTS – probability histograms, means, variances, and standard deviations of random variables, comparing probability distributions

The probability histograms consist of bars centered on the number of persons (displayed on the vertical axis and with heights equal to the corresponding probabilities). Use the axes provided below to draw the two histograms.

Family size

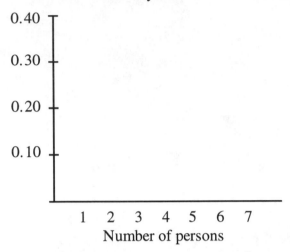

Number of persons

Household size

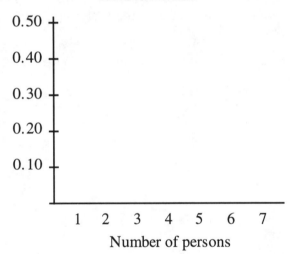

Number of persons

To calculate the mean μ_X and standard deviation σ_X, use the formulas

$$\mu_X = x_1 p_1 + x_2 p_2 + \ldots + x_k p_k$$

$$\sigma_X^2 = (x_1 - \mu_X)^2 p_1 + (x_2 - \mu_X)^2 p_2 + \ldots + (x_k - \mu_X)^2 p_k$$

Use your results to compare the two distributions.

Exercise 4.69

KEY CONCEPTS – rules for means and variances of random variables

a) Follow the same reasoning as in Exercise 4.63 for household and family size to calculate μ_X and σ_X.

b) Calculations are relatively easy using the results of (a) if we recall the formulas

$$\mu_{a + bX} = a + b\mu_X$$

$$\sigma^2_{a + bX} = b^2\sigma^2_X$$

What are a and b in this case?

c) This is just like (b). Again one must identify a and b.

COMPLETE SOLUTIONS

Exercise 4.51

We calculate

$$\mu_X = 0(0.10) + 1(0.15) + 2(0.30) + 3(0.30) + 4(0.15)$$
$$= 0 + 0.15 + 0.60 + 0.90 + 0.60 = 2.25$$

Exercise 4.57

a) Consecutive spins of a (fair) roulette wheel should be independent. Thus, the particular results of previous spins will not change the probability of any particular outcome on the next spin. On the next spin, black is just as likely as red. The gambler's reasoning that red is "hot" fails to recognize that spins are independent. It is based on a false understanding of random behavior.

b) The gambler is wrong again because he is assuming that consecutive cards are independent. This is not the case here. Initially, the deck contains 52 cards, half of which are red and half of which are black. However, each time a card is removed from the deck, the number of red and black cards remaining changes. For example, if I am dealt 5 red cards from a deck of 52 cards, the deck now contains only 47 cards, of which 21 are red and 26 are black. The probability that the next card is red is 21/47 which is less than the probability that the next card is black, which is 26/47.

Exercise 4.61

We compute

$$\sigma^2{}_X = (0 - 2.25)^2(0.10) + (1 - 2.25)^2(0.15) + (2 - 2.25)^2(0.30)$$
$$+ (3 - 2.25)^2(0.30) + (4 - 2.25)^2(0.15)$$

$$= 0.50625 + 0.234375 + 0.01875 + 0.16875 + 0.459375$$

$$= 1.3875$$

and taking the positive square root gives

$$\sigma_X = 1.1779$$

Exercise 4.63

We calculate

Household size:

$$\mu_X = 1(0.25) + 2(0.32) + 3(0.17) + 4(0.15) + 5(0.07) + 6(0.03) + 7(0.01)$$

$$= 0.25 + 0.64 + 0.51 + 0.60 + 0.35 + 0.18 + 0.07$$

$$= 2.60$$

$$\sigma^2{}_X = (1 - 2.6)^2(0.25) + (2 - 2.6)^2(0.32) + (3 - 2.6)^2(0.17)$$
$$+ (4 - 2.6)^2(0.15) + (5 - 2.6)^2(0.07) + (6 - 2.6)^2(0.03) + (7 - 2.6)^2(0.01)$$

$$= 0.64 + 0.1152 + 0.0272 + 0.294 + 0.4032 + 0.3468 + 0.1936$$

$$= 2.02$$

$$\sigma_X = \sqrt{2.02} = 1.42$$

Family size:

$$\mu_X = 1(0) + 2(0.42) + 3(0.23) + 4(0.21) + 5(0.09) + 6(0.03) + 7(0.02)$$

$$= 0 + 0.84 + 0.69 + 0.84 + 0.45 + 0.18 + 0.14$$

$$= 3.14$$

$$\sigma^2_X = (1 - 3.14)^2(0) + (2 - 3.14)^2(0.42) + (3 - 3.14)^2(0.23)$$
$$+ (4 - 3.14)^2(0.21) + (5 - 3.14)^2(0.09) + (6 - 3.14)^2(0.03)$$
$$+ (7 - 3.14)^2(0.02)$$

$$= 0 + 0.5458 + 0.0045 + 0.1553 + 0.3114 + 0.2454 + 0.2980$$

$$= 1.5604$$

$$\sigma_X = \sqrt{1.5604} = 1.2492$$

Here are the completed probability histograms.

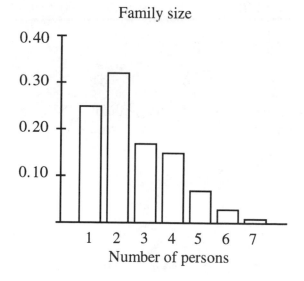

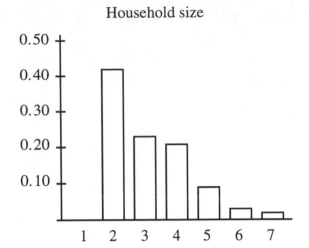

Household size

Number of persons

Household size tends to be bigger than family size by about 1. The probability histogram for household size is shifted to the right of that for family size by about 1 unit, with little increase in the variability (the standard deviations are fairly close). This is further supported by the fact that the mean household size is larger than the mean family size by a little over 1.

Exercise 4.69

a) We compute

$$\mu_X = 540(0.1) + 545(0.25) + 550(0.3) + 555(0.25) + 560(0.1)$$

$$= 54 + 136.25 + 165 + 138.75 + 56$$

$$= 550$$

$$\sigma^2_X = (540 - 550)^2(0.1) + (545 - 550)^2(0.25) + (550 - 550)^2(0.25)$$
$$+ (555 - 550)^2(0.25) + (560 - 550)^2(0.1)$$

$$= 10 + 6.25 + 0 + 6.25 + 10$$

$$= 32.5$$

$$\sigma_X = \sqrt{32.5} = 5.7$$

b) Here we are interested in $X - 550$, so $a = -550$ and $b = 1$. Thus, we have

$$\mu_{-550 + X} = -550 + 1\mu_X = -550 + 550 = 0$$

$$\sigma^2_{-550 + X} = 1^2 \sigma^2_X = 32.5$$

hence

$$\sigma_{-550 + X} = \sqrt{32.5} = 5.7$$

c) Here we want $(9/5)X + 32$, so $a = 32$ and $b = 9/5$. Thus,

$$\mu_{32 + (9/5)X} = 32 + (9/5)\mu_X = 32 + (9/5)550 = 32 + 990 = 1022$$

$$\sigma^2_{32 + (9/5)X} = (9/5)^2 \sigma^2_X = (81/25)32.5 = 105.3$$

Hence,

$$\sigma_{32 + (9/5)X} = \sqrt{105.3} = 10.26$$

SECTION 4.5

OVERVIEW

This section discusses a number of basic concepts and rules that are used to calculate probabilities of complex events. The **complement** A^c of an event A contains all the outcomes in the sample space that are not in A. It is the "opposite" of A. The **union** of two events A and B contains all outcomes in A, in B, or in both. The union is sometimes referred to as the event A or B. The **intersection** of two events A and B contains all outcomes that are in both A and B simultaneously. The interesection is sometimes referred to as the event A and B. We say that two events A and B are **disjoint** if they have no outcomes in common.

The **conditional probability** of an event B given an event A is denoted $P(B|A)$ and is defined by

$$P(B|A) = \frac{P(A \text{ and } B)}{P(A)}$$

$P(A \cap B)$

when $P(A) > 0$. In practice, it can often be determined directly from the information given in a problem. Two events A and B are **independent** if $P(B|A) = P(B)$.

Other general rules of clementary probability are

- Legitimate values: $0 \le P(A) \le 1$ for any event A

- Total probability: $P(S) = 1$, where S denotes the sample space

- Complement rule: $P(A^c) = 1 - P(A)$

- Addition rule: $P(A \text{ or } B) = P(A) + P(B) - P(A \text{ and } B)$

- Multiplication rule: $P(A \text{ and } B) = P(A)P(B|A)$

- Bayes rule: $P(A \mid B) = \dfrac{P(B \mid A)P(A)}{P(B \mid A)P(A) + P(B \mid A^c)P(A^c)}$

$$\text{provided } 0 < P(A), P(B) < 1$$

- For disjoint events: $P(A \text{ and } B) = 0$ and so $P(A \text{ or } B) = P(A) + P(B)$

- For independent events: $P(A \text{ and } B) = P(A)P(B)$

In problems with several stages, it is helpful to draw a tree diagram to guide you in the use of the multiplication and addition rules.

GUIDED SOLUTIONS

Exercise 4.75

KEY CONCEPTS – the addition rule

This is an application of the addition rule $P(A \text{ or } B) = P(A) + P(B) - P(A \text{ and } B)$.

Exercise 4.77

KEY CONCEPTS – Venn diagrams

a) From Exercise 4.75, we have the following facts about A and B.

$P(A) = 0.125$
$P(B) = 0.237$
$P(A \text{ and } B) = 0.077$
$P(A \text{ or } B) = 0.285$
$P(\text{entire sample space}) = 1$

Below is a Venn diagram showing A and B. Shade the portion representing

{A and B}.

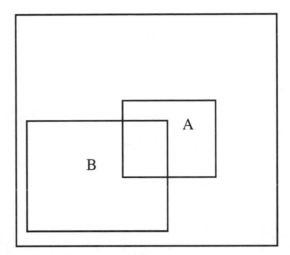

We are given that $P(A \text{ and } B) = 0.077$. Now describe in words what the event A and B means. Refer to Exercise 4.75 for the meaning of A and B.

b) Below is a Venn diagram showing A and B. Shade the portion representing {A and B^c}.

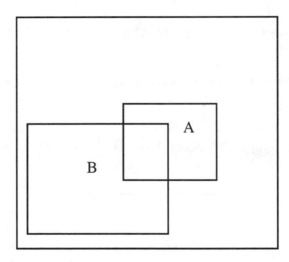

Use the facts listed in (a) and the Venn diagram to calculate $P(A \text{ and } B^c)$.

Now describe in words what the event A and B^c means. Refer to Exercise 4.75 for the meaning of A and B.

c) Below is a Venn diagram showing A and B. Shade the portion representing {A^c and B}.

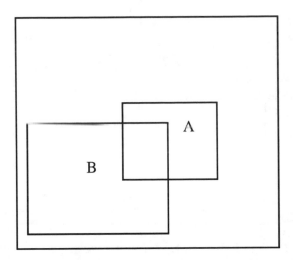

Use the facts listed in (a) and the Venn diagram to calculate $P(A^c$ and B).

Now describe in words what the event A^c and B means. Refer to Exercise 4.75 for the meaning of A and B.

d) Below is a Venn diagram showing A and B. Shade the portion representing {A^c and B^c}.

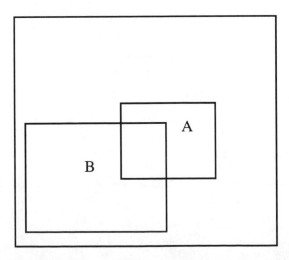

Use the facts listed in (a) and the Venn diagram to calculate $P(A^c \text{ and } B^c)$.

Now describe in words what the event A^c and B^c means. Refer to Exercise 4.75 for the meaning of A and B.

Exercise 4.79

KEY CONCEPTS – conditional probabilities and the multiplication rule

Table 4.1 is reproduced below to assist you.

	Age 18 to 24	25 to 64	65 and over	Total
Married	3046	48116	7767	58929
Never married	9289	9252	768	19309
Widowed	19	2425	8636	11080
Divorced	260	8916	1091	10267
Total	12614	68709	18262	99585

a) You can calculate this probability directly from the table. All women in the table are equally likely to be selected (that is what it means to select a woman at random), so that the fraction of the women in the table 65 years or older is the desired probability. How many women are 65 years or older? Where do you find this in the table? What is the total number of women represented in the table? Use these numbers to compute the desired fraction.

b) This probability can also be calculated directly from the table. Since this is a conditional probability (i.e., this is a probability given that the woman is 65 or over), we restrict ourselves only to women that are 65 or over. The desired probability is then the fraction of these women who are married. Find the appropriate entries in the table to compute this fraction.

c) The number of women who are both married and in the over-65 age group can be read directly from the table. What is this number? What fraction of the total number of women represented in the table is this number? This is the desired probability.

d) Recall that the multiplication rule says

    P(woman is both "married" and "65 and over")

    = P(woman is married|woman is 65 and over)P(woman is 65 and over)

Now verify that this holds, using the answers to (a), (b), and (c).

Exercise 4.91

KEY CONCEPTS – tree diagrams

We are given the following probabilities:

    P(Linda knows the answer) = 0.75

    P(Linda is correct | Linda does not know the answer) = 0.2

It also must be the case that

    P(Linda is correct | Linda knows the answer) = 1

Now fill in the probabilities for the branches in the tree diagram below and compute P(Linda is correct). Remember that the sum of the probabilities of all branches starting from the same point in the diagram is 1, assuming that all possibilities from the starting point are represented by the branches (you can verify that this is true in the tree diagram below). The probability of any endpoint in the tree is the product of the probabilities on the branches leading from the beginning of the tree to the endpoint. P(Linda is correct) is the sum of all the probabilities of endpoints labeled "Linda is correct."

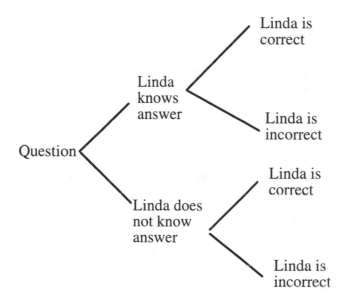

Exercise 4.93

KEY CONCEPTS – Bayes rule

What we want to calculate is

P(Linda knows answer | Linda is correct)

Note that this is not 1, since Linda might not know the answer but by good luck just guess the correct answer.

To calculate the probability, we could use Bayes rule, which says

$$P(A|B) = \frac{P(B|A)P(A)}{P(B|A)P(A) + P(B|A^c)P(A^c)}$$

For our problem, what are A and B? What are the various probabilities on the right side of this equation? (Hint: Refer to Exercise 4.91 for these probabilities.)

COMPLETE SOLUTIONS

Exercise 4.75

We are given that $P(A) = 0.125$, $P(B) = 0.237$, and PA and B) = 0.077, hence

$$P(A \text{ or } B) = 0.125 + 0.237 - 0.077 = 0.285$$

Exercise 4.77

a) The shaded area below is {A and B}.

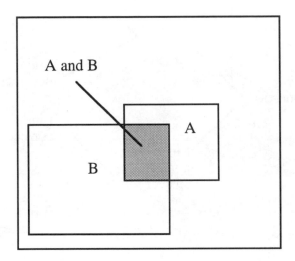

In words, the event A and B is the event that the household selected is both prosperous and educated.

b) The shaded area below is A and B^c.

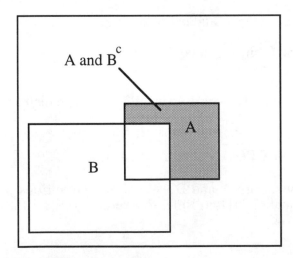

From the diagram, we see that we can write

$$P(A \text{ and } B^c) = P(A) - P(A \text{ and } B) = 0.125 - 0.077 = 0.048$$

In words, the event A and B^c is the event that the household selected is prosperous but not educated.

c) The shaded area below is { A^c and B }.

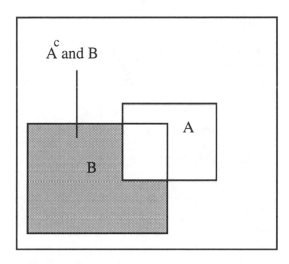

From the diagram, we see that we can write

$$P(A^c \text{ and } B) = P(B) - P(A \text{ and } B) = 0.237 - 0.077 = 0.160$$

In words, the event A^c and B is the event that the household selected is not prosperous but is educated.

d) The shaded below area is { A^c and B^c }

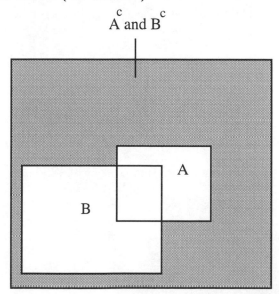

From the diagram we see that we can write

$$P(A^c \text{ and } B^c) = 1 - P(A \text{ or } B) = 1 - 0.285 = 0.715$$

In words, the event A^c and B^c means the event that the household selected is neither prosperous nor educated.

Exercise 4.79

a) The number of women 65 years or older is found in at the bottom of the column labeled "65 and over" and is (in thousands) 18262. The total number of women in the table is in the lower right corner and is (in thousands) 99585. The desired probability is thus

(number of women 65 years or older)/(total number of women in table)

$$= 18262/99585$$

$$= 0.1834$$

b) The desired probability is

(number of married women that are 65 and over)/(number of women 65 years and over)

$$= 7767/18262$$

$$= 0.4253$$

c) The number of women that are both married and in the over 65 age group is the entry in the intersection of the row labeled "Married" and the column labeled "65 and over". This number is 7767. The desired probability is

(number of women both "married" and "65 and over") / (total number of women in table)

$$= 7767/99585$$

$$= 0.0780$$

d) We have found

$P(\text{woman is both "married" and "65 and over"}) = 0.0780$

$P(\text{woman is married|woman is 65 and over}) = 0.4253$

$P(\text{woman is 65 and over}) = 0.1834$

and we see that indeed

P(woman is married|woman is 65 and over)P(woman is 65 and over)

$$= (0.4253)(0.1834)$$

$$= 0.0780$$

$$= P\text{(woman is both "married" and "65 and over")}$$

Exercise 4.91

We know

P(Linda knows the answer) $= 0.75$

P(Linda is correct | Linda does not know the answer) $= 0.20$

Our completed tree diagram looks like the following. The numbers in the boxes are those that were given in the problem.

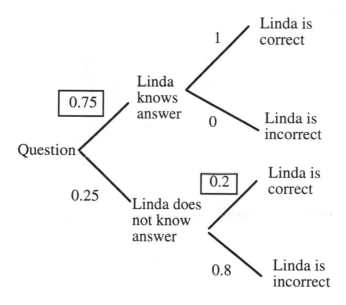

We find

P(Linda is correct) $= (0.75)(1) + (0.25)(0.2) = 0.8$

Exercise 4.93

For our problem we take

A = Linda knows answer

B = Linda is correct.

Referring to Exercise 4.91 we know

$P(B|A) = P$(Linda is correct | Linda knows answer) = 1

$P(A) = P$(Linda knows answer) = 0.75

$P(B|A^c) = P$(Linda is correct | Linda does not know answer) = 0.2

$P(A^c) = P$(Linda does not know answer) = 0.25.

Thus from Bayes rule

$$P(\text{Linda knows answer | Linda is correct}) = \frac{1 \times 0.75}{(1 \times 0.75) \; + \; (0.2 \times 0.25)} = \frac{0.75}{0.8}$$

$$= 0.9375$$

CHAPTER 5

FROM PROBABILITY
TO INFERENCE

SECTION 5.1

OVERVIEW

One of the most common situations giving rise to a **count X** is the **binomial setting**. It consists of four assumptions about how the count was produced. They are

- The number n of observations is fixed.
- The n observations are all independent.
- Tach observation falls into one of two categories called "success" and "failure."
- The probability p of success is the same for each observation.

When these assumptions are satisfied, the number of successes X has a **binomial distribution** with n trials and success probability p, denoted by $B(n, p)$. For smaller values of n, the probabilities for X can be found easily using statistical software. Table C in the text gives the probabilities for certain combinations of n and p, and there is also an exact formula. For large n, the **normal approximation** can be used.

For a large population containing a proportion p of successes, the binomial distribution is a good approximation to the number of successes in an SRS of size n, provided the population is at least 10 times larger than the sample.

The mean and standard deviation for the binomial count X and the sample proportion $\hat{p} = X/n$ can be found using the formulas

$$\mu_X = np \qquad\qquad\qquad \mu_{\hat{p}} = p$$

$$\sigma_X = \sqrt{np(1-p)} \qquad\qquad \sigma_{\hat{p}} = \sqrt{\frac{p(1-p)}{n}}$$

When n is large the count X is approximately $N(np, \sqrt{np(1-p)})$, and the proportion $\hat{p}$ is approximately $N(p, \sqrt{\frac{p(1-p)}{n}})$. These approximations should work well when $np \geq 10$ and $N(1-p) \geq 10$. For the count X, the **continuity correction** improves the approximation, particularly when the values of n and p are closer to these cutoffs.

The exact **binomial probability formula** is given by

$$P(X = k) = \binom{n}{k} p^k (1-p)^{n-k}$$

where $k = 0, 1, 2,..., n$ and $\binom{n}{k}$ is the **binomial coefficient**

GUIDED SOLUTIONS

Exercise 5.1

KEY CONCEPTS - binomial setting

There are four assumptions that must be satisfied to ensure that the count X has a binomial distribution: The number of observations or trials is fixed in advance, each trial results in one of two outcomes, the trials are independent. and the probability of success is the same from trial to trial. In addition, for a large population with a proportion p of successes, we can use the binomial distribution as an approximation to the distribution of the count X of successes in an SRS of size n. For each setting below, see if all four assumptions are satisfied.

a) Think about how this fits in the binomial setting. What is n? What are the two outcomes, and why might the trials be considered independent?

b) Think about how many observations or trials there are going to be.

c) Think about whether the trials are independent.

Exercise 5.7

KEY CONCEPTS - normal approximation for proportions

a) This question does not require computing probabilities. From the information given, what is the value of $\hat{p}$ for these data?

b) We are told that $p = 0.9$ in the population, and an SRS of 100 from the 5000 orders is taken. Since our population (5000) is more than 10 times the size of the sample (100), the number X if items in the sample shipped on time has approximately a binomial distribution with $n = 100$ trials and $p = 0.9$. You need to compute a probability about $\hat{p}$, namely $P(\hat{p} \leq 0.86)$. Since $\hat{p} \leq 0.86$ whenever $X \leq 86$, if you have statistical software that computes binomial probabilities, try to find the exact probability.

If you don't have access to statistical software that computes binomial probabilities, you'll need to use the normal approximation to the sampling distribution of $\hat{p}$ to approximate the probability. Since $np = 100(0.9) = 90$ and $n(1 - p) = 100(0.1) = 10$ are both greater than or equal to 10, we can use the approximation. (However, since n and p just meet the criteria, you might expect the normal approximation not to be that accurate.) To use the normal approximation, the mean and standard deviation of $\hat{p}$ must be computed. Do this using the formulas given below.

mean $\qquad\qquad\qquad \mu_{\hat{p}} = p =$

standard deviation $\qquad\qquad \sigma_{\hat{p}} = \sqrt{\dfrac{p(1 - p)}{n}} =$

Computing $P(\hat{p} \leq 0.86)$ is a normal probability calculation like those in 1.30. You may want to review the material there to refresh your memory on how to do such calculations, as they form the basis of solving many of the exercises in this chapter. To compute the probability, we standardize the number 0.86 (compute its z-score) by subtracting the mean $\mu_{\hat{p}}$ and dividing the result by $\sigma_{\hat{p}}$. Next, use Table A to determine the area to the left of this z-score. It may be helpful to draw a normal curve area to help you visualize the area. (You could also do the problem in terms of the count X. Both ways are illustrated in the complete solution.)

$P(\hat{p} \leq 0.86) =$

c) What does sampling variability tell us?

Exercise 5.9

KEY CONCEPTS - normal approximation for proportions

a) If $\hat{p}$ is to be within 0.03 of $p = 0.44$, then it must be between 0.41 and 0.47. So the probability of a margin of error of $\pm 3\%$ is $P(0.41 \le \hat{p} \le 0.47)$. Approximate this probability using the normal approximation to the sampling distribution of a proportion. First find the mean and standard deviation as in Exercise 5.7, and then standardize to use Table A. A picture may help.

mean = $\mu_{\hat{p}}$ =

standard deviation = $\sigma_{\hat{p}}$ =

$P(0.41 \le \hat{p} \le 0.47)$ =

b) When n increases, the mean of the distribution stays the same, but the standard deviation will change. Why does this happen? Will the standard deviation increase or decrease, and how should this affect the probability of being within $\pm 3\%$ of p? Try to answer this without doing the calculation and then carry out the computation to verify your intuition.

Exercise 5.11

KEY CONCEPTS - binomial probabilities

a) X is the number of women in the sample of 10 who have never been married. This fits the binomial setting with $n = 10$ trials, and letting "success" correspond to having never been married, we have $p = 0.25$. So X is B(10, 0.25). The probabilities below have been calculated using a statistical software package. They can also be found in Table C of the text. Once the probabilities are obtained, probabilities of events described in terms of X are found as in Chapter 4.

k	$P(X = k)$
0	0.056314
1	0.187712
2	0.281568
3	0.250282
4	0.145998
5	0.058399
6	0.016222
7	0.003090
8	0.000386
9	0.000029
10	0.000001

$P(X = 2) =$

b) Write the event "two or fewer have never been married" in terms of X and then use the tabled probabilities to find the desired probability.

c) Write the event "at least eight have been married" in terms of X and then use the tabled probabilities to find the desired probability.

Exercise 5.17

KEY CONCEPTS - binomial probabilities, mean and variance of the binomial

a) Let X = number of truthful persons "failing" the lie detector test. X has a binomial distribution with $n = 12$ and $p = 0.2$. The binomial probabilities can be found in Table C in your text or using statistical software. The probabilities below were obtained using statistical software.

k	$P(X = k)$
0	0.068719
1	0.206158
2	0.283468
3	0.236223
4	0.132876

k	P(X = k)
5	0.053150
6	0.015502
7	0.003322
8	0.000519
9	0.000058
10	0.000004
11	0.000000
12	0.000000

Use these to evaluate the probability required in (a).

b) Use the formula that expresses the mean and standard deviation of the binomial in terms of n and p.

mean =

standard deviation =

c) Don't be confused by the fact that the mean is not a whole number. Just determine which values of X are less than the mean and then add up their probabilities.

Exercise 5.19

KEY CONCEPTS - normal approximation for counts

a) When X is known to have a binomial distribution, you can use the formulas that express the mean and standard deviation of X in terms of n and p.

mean =

standard deviation =

b) This part can be done using the normal approximation for a count. First check to see that np and $n(1 - p)$ are both greater than or equal to 10. Then use the mean and standard deviation from (a) and the normal approximation to evaluate $P(X \leq 170)$.

$P(X \leq 170) =$

COMPLETE SOLUTIONS

Exercise 5.1

a) The number of trials (50) is fixed, each child is a boy or girl, whether or not a child is a boy will not alter the probability that any other children are either girls or boys (excluding identical twins to keep it simple), and the probability of any child being a boy should be the same (about 0.5) for each birth. The binomial distribution should be a good probability model for the number of girls.

b) Although each birth is a boy or girl, we are not counting the number of successes in a fixed number of births. The number of observations (births) is random. The assumption of a fixed number of observations is violated.

c) This is not the same as interviewing 50 people at random. The husband and wife may tend to share the same opinion, and the trials will not be independent of one another. If we know the wife agrees, then it is more likely that the husband agrees. The assumption of independent observations is violated.

Exercise 5.7

a) $\hat{p} = 86/100 = 0.86$.

b) Using statistical software yields $P(\hat{p} \leq 0.86) = P(X \leq 86) = 0.1239$. The normal approximation can be used to approximate the probability expressed equivalently in terms of either the proportion or the count, with the same numerical answer. We do both here and illustrate the continuity correction for the count.

For $\hat{p}$, the mean is $p = 0.9$, and the standard deviation is

$$\sqrt{\frac{p(1-p)}{n}} = \sqrt{\frac{0.9(1-0.9)}{100}} = 0.03$$

$$P(\hat{p} \leq 0.86) = P\left(\frac{\hat{p}-0.9}{0.03} \leq \frac{0.86-0.9}{0.03}\right) = P(Z \leq -1.33) = 0.0918,$$

where you need to remember that this is an approximation to $P(\hat{p} \leq 0.86)$. The agreement with the exact answer 0.1239 is not that good.

For X, the mean is $np = 90$ and the standard deviation is

$$\sqrt{np(1-p)} = \sqrt{100(0.9)(0.1)} = 3$$

$$P(X \leq 0.86) = P\left(\frac{X-90}{3} \leq \frac{86-90}{3}\right) = P(Z \leq -1.33) = 0.0918$$

where you again need to remember that this is an approximation.

For those students who have covered the continuity correction in class, we show how it can be used to improve the approximation. Since we are including the outcome 86, we include the entire interval from 85.5 to 86.5 that is the base of $X = 86$ in the histogram. So $P(X \leq 86)$ is calculated as $P(X \leq 86.5)$, giving

$$P(X \le 86.5) = P\left(\frac{X-90}{3} \le \frac{86.5-90}{3}\right) = P(Z \le -1.17) = 0.1210$$

which is much closer to the exact value of 0.1239. Remember that the continuity correction can be applied directly only to X, not to $\hat{p}$. If you want to use the continuity correction to approximate a probability about $\hat{p}$, you should first change it to a probability about X as we did here. Don't try adding 0.5 to or subtracting 0.5 from $\hat{p}$.

c) If the true percentage is 90%, the percentage $\hat{p}$ in the sample won't be exactly equal to $p = 0.9$ every time. It will fluctuate about this value according to its sampling distribution. It was shown that a value as small as or smaller than 0.86 would occur 12.39% of the time (the exact value), which is not that unusual. Our data do not contradict the claim but instead just illustrate natural sampling variability.

Exercise 5.9

a) The mean is $p = 0.44$, and the standard deviation is

$$\sqrt{\frac{p(1-p)}{n}} = \sqrt{\frac{0.44(1-0.44)}{300}} = 0.0287$$

$P(0.41 \le \hat{p} \le 0.47)$

$$= P\left(\frac{0.41-0.44}{0.0287} \le \frac{\hat{p}-0.44}{0.0287} \le \frac{0.47-0.44}{0.0287}\right) = P(-1.05 \le Z \le 1.05)$$

$$= 0.8531 - 0.1469 = 0.7062$$

b) As the sample size increases, the mean remains $p = 0.44$ and the standard deviation decreases. This makes the distribution more concentrated about its mean, so the probability of being within 3% should increase. For $n = 600$, the standard deviation is

$$\sqrt{\frac{p(1-p)}{n}} = \sqrt{\frac{0.44(1-0.44)}{600}} = 0.0203$$

and

$P(0.41 \le \hat{p} \le 0.47)$

$$= P\left(\frac{0.41-0.44}{0.0203} \le \frac{\hat{p}-0.44}{0.0203} \le \frac{0.47-0.44}{0.0203}\right) = P(-1.48 \le Z \le 1.48)$$

$$= 0.9306 - 0.0694 = 0.8612$$

For $n = 1200$ the standard deviation is $\sqrt{\dfrac{p(1-p)}{n}} = \sqrt{\dfrac{0.44(1-0.44)}{600}} = 0.0143$

and $P(\,0.41 \le \hat{p} \le 0.47) =$

$$P\left(\dfrac{0.41-0.44}{0.0143} \le \dfrac{\hat{p}-0.44}{0.0143} \le \dfrac{0.47-0.44}{0.0143}\right) = P(-2.10 \le Z \le 2.10)$$

$$= 0.9821 - 0.0179 = 0.9642.$$

Exercise 5.11

a) $P(\,X = 2) = 0.281568$

b) The event "two or fewer have never been married" is $X \le 2$.
$P(\,X \le 2) = 0.056314 + 0.187712 + 0.281568 = 0.525594.$

c) The event "at least 8 have been married" corresponds to "two or fewer have never been married" since

8 have been married   = 2 have never been married
9 have been married   = 1 have never been married
10 have been married   = 0 have never been married.

So the probability is the same as in (b). The same event has been described in two different ways.

Exercise 5.17

a) X is B(12, 0.2). You are asked to evaluate $P(\,X \ge 1)$, the probability that the polygraph says that at least one person telling the truth is deceptive. Using the table of probabilities

k	$P(\,X = k)$
0	0.068719
1	0.206158
2	0.283468
3	0.236223
4	0.132876
5	0.053150
6	0.015502
7	0.003322
8	0.000519
9	0.000058
10	0.000004
11	0.000000
12	0.000000

you may first think to add the probabilities for $k = 1, 2, 3,..., 12$. Although this will give the correct answer, in this case it is much simpler to use the rule for complements: $P(X \geq 1) = 1 - P(X = 0) = 1 - 0.068719 = 0.931281$. When determining the complementary event, you must be careful whether an event is of the form greater than or greater than or equal to. The complement of $X \geq 2$ is $X \leq 1$, whereas the complement of $X > 2$ is $X \leq 2$.

b) The mean is $np = 12(0.2) = 2.4$, and the standard deviation is found using the formula $\sqrt{np(1-p)} = \sqrt{12(0.2)(0.8)} = 1.386$.

c) X is less than the mean 2.4 only if X is 0, 1, or 2. So, we need to find

$$P(X \leq 2) = 0.068719 + 0.206158 + 0.283468 = 0.558345$$

Exercise 5.19

a) X has a binomial distribution with $n = 1500$ and $p = 0.12$. The mean of X is np $= 1500(0.12) = 180$, and the standard deviation of X is

$$\sqrt{np(1-p)} = \sqrt{1500(0.12)(0.88)} = 12.586$$

b) The normal approximation can be used since both $np = 180$ and $n(1-p) = 1320$ are greater than 10. Using the mean and variance evaluated in (a) gives the approximation

$$P(X \leq 170) = P\left(\frac{X-180}{12.586} \leq \frac{170-180}{12.586}\right) = P(Z \leq -0.79) = 0.2148$$

If the continuity correction is used then we are including the outcome 170, and so we include the entire interval from 169.5 to 170.5 which is the base of $X = 170$ in the histogram. So $P(X \leq 170)$ is calculated as $P(X \leq 170.5)$, giving

$$P(X \leq 170.5) = P\left(\frac{X-180}{12.586} \leq \frac{170.5-180}{12.586}\right) = P(Z \leq -0.75) = 0.2266$$

The use of the continuity correction makes less of a difference in the final answer, because the condition on n and p is more than adequately met for the normal approximation to work.

SECTION 5.2

OVERVIEW

This section examines properties of the **sample mean** $\bar{x}$. If we select an SRS of size n from a large population with mean μ and standard deviation σ, the sample mean $\bar{x}$ has a sampling distribution with

$$\text{mean} = \mu_{\bar{x}} = \mu$$

and

$$\text{standard deviation} = \sigma_{\bar{x}} = \frac{\sigma}{\sqrt{n}}$$

This implies that the sample mean is an unbiased estimator of the population mean and is less variable than a single observation.

Linear combinations (such as sums or means) of independent normal random variables have normal distributions. In particular, if the population has a normal distribution, the sampling distribution of $\bar{x}$ is normal. Even if the population does not have a normal distribution, for large sample sizes the sampling distribution of $\bar{x}$ computed from an SRS is approximately normal. In particular, the **central limit theorem** states that for large n, the sampling distribution of $\bar{x}$ computed from an SRS is approximately $N(\mu, \frac{\sigma}{\sqrt{n}})$ for any population with mean μ and finite standard deviation σ.

GUIDED SOLUTIONS

EXERCISE 5.27

KEY CONCEPTS - the sampling distribution of the sample mean, normal probability calculations

a) Recall that the sampling distribution of a sample mean of observations from a normal population with mean μ and standard deviation σ is normal with

$$\text{mean} = \mu_{\bar{x}} = \mu$$

and

$$\text{standard deviation} = \sigma_{\bar{x}} = \frac{\sigma}{\sqrt{n}}$$

In this case, what are the mean μ and standard deviation σ of the population? What is n, the sample size?

b) This is just a normal probability calculation, like those we did in Section 1.3. You may wish to review the material there to refresh your memory as to how to do such calculations. We want the probability that the sample mean $\bar{x}$ is 124 mg or higher. To compute this probability, we standardize the number 124 (compute its z-score) by subtracting the mean $\mu_{\bar{x}}$ and dividing the result by $\sigma_{\bar{x}}$. You computed $\mu_{\bar{x}}$ and $\sigma_{\bar{x}}$ in (a). Next use Table A to determine the area to the right of this z-score under a standard normal curve. You may wish to draw a picture of the standard normal to help you visualize the desired area.

EXERCISE 5.31

KEY CONCEPTS - means and standard deviations of random variables, the law of large numbers, the central limit theorem, the sampling distribution of the sample mean, normal probability calculations

a) You may wish to review how to calculate the mean and standard deviation of a random variable. This was discussed in Chapter 4. Recall that if X is a discrete random variable having possible values $x_1, x_2,..., x_k$ with corresponding probabilities $p_1, p_2,..., p_k$, the mean μ_X is the average of the possible values weighted by the corresponding probabilities, i.e.,

$$\mu_X = x_1p_1 + x_2p_2 + ... + x_kp_k$$

The variance is

$$\sigma^2_X = (x_1 - \mu_X)^2p_1 + (x_2 - \mu_X)^2p_2 + ... + (x_k - \mu_X)^2p_k$$

and the standard deviation σ_X is the positive square root of the variance.

For simplicity, assume that the gambler bets on red. What is the probability he will win, assuming all slots are equally likely to contain the ball? Complete the table below to help you compute the mean and standard deviation of the outcomes.

Outcome (winnings) X	$1	−$1
Probability		

$\mu_X =$

$\sigma^2_X =$

$\sigma_X =$

b) Recall that the law of large numbers (see Chapter 4) tells us that the average of the values of X observed in many trials must approach μ_X. Interpret this in the context of this problem, using plain English.

c) The central limit theorem states that for large n, the sampling distribution of the mean winnings computed from an SRS is approximately $N(\mu_X, \frac{\sigma_X}{\sqrt{n}})$. What are n, μ_X, and σ_X here?

Refer to Chapter 1 to refresh your memory about the 68–95–99.7 rule. Remember to use μ_X for the mean and $\frac{\sigma_X}{\sqrt{n}}$ for the standard deviation when applying the rule.

d) From (c) we found that the sampling distribution of the gambler's mean winnings $\bar{x}$ if he makes 50 bets is $N(\mu_X = -0.0526, \frac{\sigma_X}{\sqrt{n}} = 0.1412)$. We want to compute the probability that $\bar{x}$ is less than 0. This is just a normal probability calculation of the sort we studied in Section 1.30. Find the z-score of 0 and then use Table A to determine the area under the standard normal curve below this z-score.

e) Now repeat the same sorts of calculations as in (c) and (d), but with $n = 100,000$ rather than $n = 50$.

EXERCISE 5.37

KEY CONCEPTS - the sampling distribution of the sample mean, normal probability calculations

We are told that the distribution of individual scores at Southwark Elementary School is approximately normal with mean $\mu = 13.6$ and standard deviation $\sigma = 3.1$. To find L, Mr. Lavin needs to first determine the sampling distribution of the mean score $\bar{x}$ of $n = 22$ children. What is this sampling distribution? Refer to (a) of Exercise 5.27 if you need a hint.

To complete the problem, you need to find L such that the probability of $\bar{x}$ being below L is only 0.05. We did this sort of problem in Section 1.30. First refer to Table A to find the value z such that the area to the left of z under a standard normal curve is 0.05. What is this value?

This value z is the z-score of L. This means that $z = (L - \mu_{\bar{x}})/\sigma_{\bar{x}}$, where $\mu_{\bar{x}}$ and $\sigma_{\bar{x}}$ are the mean and standard deviation, respectively, of the sampling distribution of $\bar{x}$. Solve this equation for L.

EXERCISE 5.39

KEY CONCEPTS - linear combinations of independent random variables

a) We need to recall several facts from Section 4.4. First, recall that for a random variable X,

$$\mu_{a + bX} = a + b\mu_X$$

Using $a = 0$ and $b = -1$, this implies that $\mu_{-X} = -\mu_X$. Second, recall that for random variables X and Y,

$$\mu_{Y+X} = \mu_Y + \mu_X$$

Replacing X by the random variable $-X$ and using $\mu_{-X} = -\mu_X$, we have

$$\mu_{Y-X} = \mu_Y - \mu_X$$

Finally, recall that if X and Y are independent random variables, then

$$\sigma^2{}_{Y-X} = \sigma^2{}_Y + \sigma^2{}_X$$

Wc now apply these facts to $\bar{x}$ and $\bar{y}$ to get

$$\mu_{\bar{y}-\bar{x}} = \mu_{\bar{y}} - \mu_{\bar{x}}$$

and

$$\sigma^2_{\bar{y}-\bar{x}} = \sigma^2_{\bar{y}} + \sigma^2_{\bar{x}}$$

We take the square root of $\sigma^2_{\bar{y}-\bar{x}}$ to get the standard deviation of $\bar{y} - \bar{x}$. To complete the problem, what are $\mu_{\bar{y}}$, $\mu_{\bar{x}}$, $\sigma^2_{\bar{y}}$, and $\sigma^2_{\bar{x}}$ here?

b) We know that since weight gains are normally distributed in both populations, $\bar{x}$ and $\bar{y}$ are normally distributed. You determined the mean and standard deviations of the sampling distributions of these means in (a). Since we can assume that $\bar{x}$ and $\bar{y}$ are independent, we can assume that the sampling distribution of $\bar{y} - \bar{x}$ is also normal. You determined the mean and standard deviation of the sampling distribution of $\bar{y} - \bar{x}$ in (a).

c) Here we want to determine the probability that $\bar{y} - \bar{x}$ is greater than 25. We found that the sampling distribution of $\bar{y} - \bar{x}$ is $N(\mu_{\bar{y}-\bar{x}} = 25, \sigma_{\bar{y}-\bar{x}} = 16.62)$ in (b). The desired probability can be computed using the methods discussed in Section 1.3 for calculating normal probabilities.

COMPLETE SOLUTIONS

EXERCISE 5.27

a) We are given that the population has mean $\mu = 123$ mg, has standard deviation $\sigma = 0.08$ mg, and that the population is normal. Thus, we know that the sampling distribution of this mean will also be normal. In addition, the sampling distribution will have

$$\text{mean} = \mu_{\bar{x}} = \mu = 123 \text{ mg}$$

and

$$\text{standard deviation} = \sigma_{\bar{x}} = \frac{\sigma}{\sqrt{n}} = \frac{0.08}{\sqrt{3}} = 0.046$$

b) Recall that $\mu_{\bar{x}} = 123$ and $\sigma_{\bar{x}} = 0.046$. The z-score of 124 is thus

$$z\text{-score} = \frac{124 - 123}{0.046} = 21.74$$

The area to the right of this under a standard normal curve is not in Table A since the z-score exceeds the values given in the Table. It is essentially 0.

EXERCISE 5.31

a) Since 18 of the 38 slots are red, the probability of winning (if you bet on red) is 18/38. Thus, we have

Outcome (winnings) X	$1	−$1
Probability	18/38	20/38

and so

$$\text{mean winnings} = \mu_X = (\$1)(18/38) + (-\$1)(20/38) = -\$2/38 = -\$0.0526$$

$$\text{variance of winnings (in dollars squared)} = \sigma^2_X$$
$$= (1 - [-2/38])^2(18/38) + (-1 - [-2/38])^2(20/38)$$
$$= 0.5249 + 0.4724$$
$$= 0.9973$$

$$\text{standard deviation of winnings} = \sigma_X = \$\sqrt{0.9973} = \$\,0.9986$$

b) The law of large numbers tells us that if a gambler makes a large number of bets on red, his mean winnings per bet will be approximately $\mu_X = -\$0.0526$. In other words, he will lose $0.0526 on the average per bet.

c) Here $n = 50$, $\mu_X = -\$0.0526$, and $\sigma_X = \$0.9986$. The approximate distribution of the gambler's mean winnings in 50 bets is thus normal with

$$\text{mean} = \mu_X = -\$0.0526$$

and

$$\text{standard deviation} = \frac{\sigma_X}{\sqrt{n}} = \frac{\$0.9986}{\sqrt{50}} = \$0.1412$$

The 68–95–99.7 rule says that the middle 95% of the gambler's mean winnings on nights when he places 50 bets is within two standard deviations of the mean. In other words, μ_X lies between $\mu_X \pm 2\frac{\sigma_X}{\sqrt{n}} = -\$0.0526 \pm 2(\$0.1412)$ or between and $-\$0.3350$ and $\$0.2298$. Multiplying by 50 to convert to total winnings, we get that the middle 95% of the gambler's total winnings on nights when he places 50 bets is between $-\$16.75$ and $\$11.49$.

d) The sampling distribution of the gambler's mean winnings $\bar{x}$ if he makes 50 bets is $N(\mu_X = -0.0526, \frac{\sigma_X}{\sqrt{n}} = 0.1412)$. To find the probability that $\bar{x}$ is less than 0, we compute the z-score of 0, which is

$$z\text{-score} = (0 - [-0.0526])/(0.1412) = 0.0526/0.1412 = 0.37$$

According to Table A, the area under the standard normal curve to the left of 0.37 is 0.6443. Thus, the probability that the gambler will lose money if he makes 50 bets is 0.6443.

e) If $n = 100,000$ bets are made on red, the sampling distribution of the mean winnings is again normal, but now with

$$\text{mean} = \mu_X = -\$0.0526$$

and

$$\text{standard deviation} = \frac{\sigma_X}{\sqrt{n}} = \frac{\$0.9986}{\sqrt{100,000}} = \$0.00316$$

The 68–95–99.7 rule says that the middle 95% of the mean winnings of gamblers on these 100,000 bets is within two standard deviations of the mean. In other words, lies between $\mu_X \pm 2\dfrac{\sigma_X}{\sqrt{n}} = -\$0.0526 \pm 2(\$0.00316)$ or between $-\$0.05892$ and $-\$0.04628$. Multiplying by 100,000 to convert to total winnings, we get that the middle 95% of the mean winnings on 100,000 bets is between $-\$5892$ and $-\$4628$.

EXERCISE 5.37

The sampling distribution of the mean score $\bar{x}$ of 22 children is approximately normal with mean

$$\mu_{\bar{x}} = \mu = 13.6$$

and standard deviation

$$\sigma_{\bar{x}} = \frac{\sigma}{\sqrt{n}} = \frac{3.1}{\sqrt{22}} = 0.66$$

Next, we note from Table A that the value of z such that the area to the left of it under a standard normal curve is 0.05 is $z = -1.65$. Thus

$$-1.65 = (L - 13.6)/0.66.$$

Solving for L gives

$$L = (-1.65)(0.66) + 13.6 = 12.51$$

EXERCISE 5.39

a) We know that the mean and standard deviation of X are 360 g and 55 g and the mean and standard deviation of Y are 385 g and 50 g. For an average based on a sample size of $n = 20$, we have

$$\mu_{\bar{x}} = 360$$

$$\sigma_{\bar{x}}^2 = (55)^2/(20) = 151.25$$

$$\mu_{\bar{y}} = 385$$

$$\sigma_{\bar{y}}^2 = (50)^2/(20) = 125$$

Therefore

$$\mu_{\bar{y}-\bar{x}} = \mu_{\bar{y}} - \mu_{\bar{x}} = 385 - 360 = 25$$

$$\sigma_{\bar{y}-\bar{x}}^2 = \sigma_{\bar{y}}^2 + \sigma_{\bar{x}}^2 = 151.25 + 125 = 276.25$$

$$\sigma_{\bar{y}-\bar{x}} = \sqrt{276.25} = 16.62$$

b) From (a), we have

$$\mu_{\bar{x}} = 360$$

$$\sigma_{\bar{x}}^2 = (55)^2/(20) = 151.25 \text{ (hence } \sigma_{\bar{x}} = \sqrt{151.25} = 12.30)$$

$$\mu_{\bar{y}} = 385$$

$$\sigma_{\bar{y}}^2 = (50)^2/(20) = 125 \text{ (hence } \sigma_{\bar{y}} = \sqrt{125} = 11.18)$$

$$\mu_{\bar{y}-\bar{x}} = 25$$

$$\sigma_{\bar{y}-\bar{x}} = 16.62$$

Thus,

the distribution of $\bar{x}$ is $N(\mu_{\bar{x}} = 360, \sigma_{\bar{x}} = 12.30)$

the distribution of $\bar{y}$ is $N(\mu_{\bar{y}} = 385, \sigma_{\bar{y}} = 11.18)$

the distribution of $\bar{y} - \bar{x}$ is $N(\mu_{\bar{y}-\bar{x}} = 25, \sigma_{\bar{y}-\bar{x}} = 16.62)$.

c) To determine the probability that $\bar{y} - \bar{x}$ is greater than 25, we compute the z-score of 25. We get

z-score $= (25 - \mu_{\bar{y}-\bar{x}})/\sigma_{\bar{y}-\bar{x}} = (25 - 25)/16.62 = 0$.

The area to the right of 0 under a standard normal curve is 0.5, so this is the probability that $\bar{y} - \bar{x}$ is greater than 25.

SECTION 5.3

OVERVIEW

This section examines statistical methodology, called **statistical process control**, for monitoring a process over time so that any changes in the process can be detected and corrected quickly. This is an economical method for maintaining product quality in a manufacturing process. We say that a process that continues over time is in **control** if it is operating under stable conditions. More precisely, the process in is control with respect to some variable measured on the process if the distribution of this variable remains constant over time.

Control charts are used to monitor the values of a variable measured on a process over time. One of the most common control charts is the $\bar{x}$ **control chart**. This is produced by observing a sample of n values of the variable of interest periodically and plotting the means of these values versus the time order of the samples on a graph. A solid **centerline** at the target value of the process mean m for the variable is drawn on the graph, as are dashed **control limits** at $\mu \pm 3\dfrac{\sigma}{\sqrt{n}}$, where σ is the process standard deviation of the variable.

This chart helps us decide if the process is in control with mean μ and standard deviation σ. The probability that the next point (value of $\bar{x}$) on such a chart lies outside the control limits is about 0.003 if the process is in control. Such a point would be evidence that the process is **out of control**, that is, that the distribution of the process has changed for some reason. When a process is deemed out of control, a cause for the change in the process should be sought.

In addition to a point plotting outside the control limits, there are other signals that suggest the process is out of control. One is a **run** of 9 consecutive points on the same side of the centerline. Whatever signal we use, we can compute the probability that a process that is in control will give the signal. If this probability is small, we can regard the signal as evidence that the process is out of control.

There are many other types of control charts, and they are usually named after the statistic that is plotted on the chart. For example, the $\hat{p}$ **chart** plots the proportion of times a certain characteristic is observed in samples of size n versus the time order of the samples.

SAMPLE PROBLEMS

GUIDED SOLUTIONS

EXERCISE 5.51

KEY CONCEPTS - $\bar{x}$ control charts

a) For an $\bar{x}$ control chart, we plot the values of $\bar{x}$ in time order. In this case, lot number corresponds to time order. On the control chart we also plot a solid centerline at the target value or standard value of the process mean m and dashed control limits at $\mu \pm 3\dfrac{\sigma}{\sqrt{n}}$, where σ is the process standard deviation.

What are the values of μ, σ, and n here? You may use the axes below to construct the control chart.

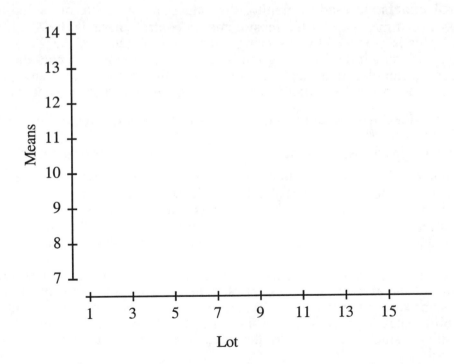

b) The "one point out rule" states that we declare the process out of control if a point plots outside the control limits. In this case, a mean breaking strength of greater than 10 is acceptable, so we only look for points below the lower control limit. Are there any such points in your control chart?

The "run of nine" rule says that we declare the process out of control if nine consecutive points on the same side of the centerline. Are there any such runs in your control chart?

What action do you recommend?

EXERCISE 5.59

KEY CONCEPTS - $\hat{p}$ charts

a) Recall that the mean and standard deviation of the sampling distribution of $\hat{p}$ are

$$\mu_{\hat{p}} = p$$

and

$$\sigma_{\hat{p}} = \sqrt{\frac{p(1-p)}{n}}$$

Compute these quantities. What is n here? What is p when the process is in control?

b) According to the normal approximation, the sampling distribution of $\hat{p}$ is normal with mean $\mu_{\hat{p}}$ and standard deviation $\sigma_{\hat{p}}$. In (a) we computed these values to be

$$\mu_{\hat{p}} = p = 0.1$$

$$\sigma_{\hat{p}} = \sqrt{\frac{p(1-p)}{n}} = \sqrt{\frac{(0.1)(0.9)}{400}} = 0.015.$$

c) The sample proportion $\hat{p}$ is just a special case of a mean. Thus the centerline and control limits will correspond to those for means, namely

centerline = $\mu_{\hat{p}}$

control limits = $\mu_{\hat{p}} \pm 3\sigma_{\hat{p}}$

Now use the results of (a) to complete the calculations.

d) Use the axes below to make your plot.

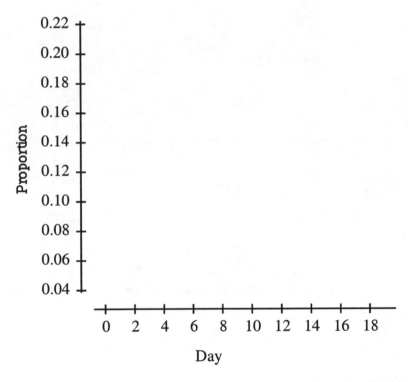

Use the "one point out rule" and "run of nine" to decide if points indicate the process is out of control.

COMPLETE SOLUTIONS

EXERCISE 5.51

We are told that $n = 3$, $\mu = 10$, and $\sigma = 1.2$. Thus, our center line is at 10 on the vertical axis and our control limits are at

$$\mu \pm 3\frac{\sigma}{\sqrt{n}} = 10 \pm 3\frac{1.2}{\sqrt{3}} = 10 \pm 2.08$$

Our completed control chart is

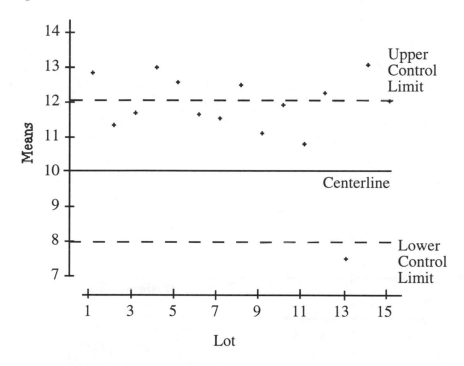

b) Examining our control chart, we notice that point 13 plots below the lower control limit, indicating that the process is out of control. We also see that the first 12 points all plot above the centerline. Thus, according to the "run of nine" rule, we would declare the process out of control.

The appropriate action would be to stop the process, try to determine why the process is out of control at lot 13, and take corrective action.

EXERCISE 5.59

a) In this problem, $n = 400$. When the process is in control, $p = 0.1$. Thus

$$\mu_{\hat{p}} = p = 0.1$$

$$\sigma_{\hat{p}} = \sqrt{\frac{p(1-p)}{n}} = \sqrt{\frac{(0.1)(0.9)}{400}} = 0.015$$

b) A complete solution is given in the guided solution.

c) We compute

centerline = $\mu_{\hat{p}}$ = 0.1

control limits = $\mu_{\hat{p}} \pm 3\sigma_{\hat{p}}$ = 0.1 ± 3(0.015) = 0.1 ± 0.045

d) The control chart looks like the following.

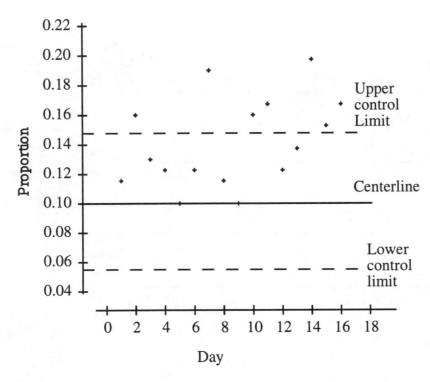

Several points plot above the upper control limit. Also, every point plots at or above the centerline. There is considerable evidence that the process is out of control.

CHAPTER 6

INTRODUCTION TO INFERENCE

SECTION 6.1

OVERVIEW

A **confidence interval** provides an estimate of an unknown parameter of a population or process along with an indication of how accurate this estimate is and how **confident** we are that the interval is correct. Confidence intervals have two parts. One is an interval computed from our data. This interval typically has the form

$$\text{estimate} \pm \text{margin of error}$$

The other part is the **confidence level**, which states the probability that the *method* used to construct the interval will give a correct answer. For example, if you use a 95% confidence interval repeatedly, in the long run 95% of the intervals you construct will contain the correct parameter value. Of course, when you apply the method only once, you do not know if your interval gives a correct value or not. Confidence refers to the probability that the method gives a correct answer in repeated use, not the correctness of any particular interval we compute from data.

Suppose we wish to estimate the unknown mean μ of a normal population with known standard deviation σ based on an SRS of size n. A level C confidence interval for μ is

$$\bar{x} \pm z^* \frac{\sigma}{\sqrt{n}}$$

where z^* is such that the probability is C that a standard normal random variable lies between $-z^*$ and z^* and z^* is obtained from the bottom row in Table D.

The margin of error $z^* \dfrac{\sigma}{\sqrt{n}}$ of a confidence interval decreases when any of the following occur.

- The confidence level C decreases.

- The sample size n increases.

- The population standard deviation σ decreases.

The sample size needed to obtain a confidence interval for a normal mean of the form

$$\text{estimate} \pm \text{margin of error}$$

with a specified margin of error m is

$$n = \left(\frac{z^* \sigma}{m} \right)^2$$

where z^* is the critical point for the desired level of confidence. Many times the n you will find will not be an integer. If it is not, round up to the next larger integer.

The formula for any specific confidence interval is a recipe that is correct under specific conditions. The most important conditions concern the methods used to produce the data. Many methods (including those discussed in this section) assume that our data were collected by random sampling. Other conditions, such as the actual distribution of the population, are also important.

GUIDED SOLUTIONS

Exercise 6.1

KEY CONCEPTS - the sampling distribution of $\bar{x}$, confidence intervals

a) Since we are willing to assume that the $n = 24$ weights are a random sample, we know that the sampling distribution of $\bar{x}$ has standard deviation $\sigma_{\bar{x}} = \dfrac{\sigma}{\sqrt{n}}$. Use the values of σ and n to compute $\sigma_{\bar{x}}$.

b) The general formula is $\bar{x} \pm z^* \dfrac{\sigma}{\sqrt{n}}$ where z^* is such that the probability is C that a standard normal random variable lies between $-z^*$ and z^* and z^* is obtained from the bottom row in Table D. To use this formula for a 95% confidence interval, one must compute

  $\bar{x}$ = mean of the 24 weights given (this is easily done using a statistical software package)

  z^* = a number such that the probability is 0.95 that a standard normal random variable lies between $-z^*$ and z^* (refer to Table D)

  $\dfrac{\sigma}{\sqrt{n}}$, which you computed in (a)

Determine each of these quantities and then substitute their values into the general formula. Where is the value 65 relative to your interval?

Exercise 6.11

KEY CONCEPTS – confidence intervals, the sample size required to obtain a confidence interval of specified margin of error

a) The margin of error of a level C confidence interval is $z^* \dfrac{\sigma}{\sqrt{n}}$, where z^* is such that the probability is C that a standard normal random variable lies between $-z^*$ and z^* and is obtained from the bottom row in Table D. To do this exercise, you must identify

C = the level of confidence required =

z^* = the probability is C that a standard normal random variable lies between $-z^*$ and z^* =

σ = population standard deviation =

n = the sample size used =

Determine the above values and then compute the margin of error

$$m = z^* \frac{\sigma}{\sqrt{n}} =$$

b) The calculations here are the same as in (a) except that n has changed from 100 to 10. Should the margin of error be larger or smaller? Why? Repeat the calculations you did in (a) with this change and see if your intuition was correct.

c) The smallest value of n that will yield a 95% confidence interval with a margin of error = m must satisfy

$$n = \left(\frac{z^* \sigma}{m} \right)^2$$

where z^* and σ are as in (a). Identify m in this case and use the above formula to compute n. Remember to round your answer up to the nearest integer. Is n within the limits of your budget as indicated in (a)?

Exercise 6.19

KEY CONCEPTS - confidence levels for several confidence intervals simultaneously, binomial probability calculations

We know the following.

- we are interested in a fixed number of intervals (seven, to be precise)

- the seven intervals are all independent

- either an interval contains the true mean yield (success) or it does not (failure)

- the probability that any particular interval will contain the true mean yield is 0.95

Let X be the number of intervals (out of the seven) that contain the true mean yield (i.e., the number of successes). X should remind you of a special type of random variable whose probability distribution we have studied previously (Hint: Look at Section 5.1). How do you calculate probabilities for X ?

a) We want the probability of seven successes. What is this probability?

b) We want the probability of at least six successes. What is this probability?

Exercise 6.23

KEY CONCEPTS - interpreting confidence intervals

Recall that the confidence level of a confidence interval states the probability that the *method* used to construct the interval will give a correct value of the population parameter being estimated. For example, if you use a 95% confidence interval repeatedly, in the long run 95% of the intervals you construct will contain the true parameter value. What parameter is being estimated in this problem? Does the student's statement convey the proper meaning of a 95% confidence interval here?

COMPLETE SOLUTIONS

Exercise 6.1

a) We are given that $\sigma = 4.5$ and $n = 24$. Hence $\sigma_{\bar{x}} = \dfrac{\sigma}{\sqrt{n}} = \dfrac{4.5}{\sqrt{24}} = 0.92$.

b) We compute $\bar{x} = 61.79$, $z^* = 1.96$, and from (a) $\dfrac{\sigma}{\sqrt{n}}, = 0.92$.

Hence

$$\bar{x} \pm z^* \frac{\sigma}{\sqrt{n}} = 61.79 \pm (1.96)(0.92) = 61.79 \pm 1.80$$

or, equivalently, 59.99 to 63.59. Since the entire interval lies below 65, we are at least 95% confident that the average weight in the population is less than 65. We can be fairly sure that the average weight of the population of runners is less than 65 kg.

Exercise 6.11

a) From the statement of the problem, we see that

C = the level of confidence required = 0.95

hence

z^* = the probability is 0.95 that a standard normal random variable lies
between $-z^*$ and z^*

= 1.96 (see Table D)

We also see that

σ = population standard deviation = 12

n = the sample size used = 100

Thus, the margin of error m is

$$m = z^* \frac{\sigma}{\sqrt{n}} = 1.96 \frac{12}{\sqrt{100}} = 2.352$$

b) We change n to 10 in (a) and find that the margin of error m is now

$$m = z^* \frac{\sigma}{\sqrt{n}} = 1.96 \frac{12}{\sqrt{10}} = 7.438.$$

c) Again we have $z^* = 1.96$ and $\sigma = 12$. We are told we want $m \leq 3$, so we use the   maximum possible value we would tolerate, namely $m = 3$, in the formula for n and get

$$n = \left(\frac{z^* \sigma}{m}\right)^2 = \left(\frac{(1.96)(12)}{3}\right)^2 = (7.84)^2 = 61.47.$$

Rounding up, we see that the smallest value of n that will accomplish our goal is $n = 62$. Since the budget will allow up to 100 students, this is within the limits of the budget.

Exercise 6.19

Referring to Section 1 of Chapter 5 we see that X has a binomial $B(n=7, p=0.95)$ distribution. We use this distribution to calculate the required probabilities. Table C in your text gives binomial probabilities.

a) Here we want the probability that $X = 7$. Table C only gives binomial probabilities for $p \leq 0.50$. Thus to use Table C we have to rewrite the desired probability in terms of the number of failures (which has $p = 0.05$, a value that is given in Table C) We have (look in the portion of the table with $n = 7$ and $p = 0.05$)

$P(X = 7) = P(\text{number of failures} = 0) = 0.6983$

b) Again, rewriting the desired probability (at least 6 successes) in terms of number of failures we have

$P(X \geq 6) = P(\text{number of failures} \leq 1) = P(0 \text{ failures}) + P(1 \text{ failure})$

$= 0.6983 + 0.2573 = 0.9556.$

Exercise 6.23

The student is not correct. The 95% confidence interval is not a statement about the proportion of the population that falls in the interval. The 95% confidence interval is a statement about the **mean** SAT math score of California high school seniors. The 95% tells us the probability that the *method* used to construct the interval will produce an interval containing the true value of the **mean** SAT math score of California high school seniors. We can be 95% confident that if we knew all the SAT math scores of California high school seniors and computed the mean of all these scores, the interval 452 to 470 would contain this mean.

SECTION 6.2

OVERVIEW

Tests of significance and confidence intervals are the two most widely used types of formal statistical inference. A test of significance is done to assess the evidence against the **null hypothesis** H_0 in favor of an **alternative hypothesis** H_a. Typically, the alternative hypothesis is the effect that the researcher is trying to demonstrate, and the null hypothesis is a statement that the effect is not present. The alternative hypothesis can be either **one-sided** or **two-sided**.

Tests are usually carried out by first computing a **test statistic**. The test statistic is used to compute a *P*-value, which is the probability of getting a test statistic at least as extreme as the one observed, where the probability is computed when the null hypothesis is true. The *P*-value provides a measure of how incompatible our data are with the null hypothesis, or how unusual it would be to get data like ours if the null hypothesis were true. Since small *P*-values indicate data that are unusual or difficult to explain under the null hypothesis, we typically reject the null hypothesis in these cases. In this case, the alternative hypothesis provides a better explanation for our data.

Significance tests of the null hypothesis H_0: $\mu = \mu_0$ with either a one-or two-sided alternative are based on the test statistic

$$z = \frac{\bar{x} - \mu_0}{\sigma / \sqrt{n}}$$

The use of this test statistic assumes that we have an SRS from a normal population with known standard deviation σ. When the sample size is large, the assumption of normality is less critical because the sampling distribution of $\bar{x}$ is approximately normal. *P*-values for the test based on z are computed using Table A.

When the *P*-value is below a specified value α, we say that the results are statistically significant at level α, or we reject the null hypothesis at level α.

Tests can be carried out at a fixed significance level by obtaining the appropriate critical value z^* from the bottom row in Table D.

GUIDED SOLUTIONS

Exercise 6.27

KEY CONCEPTS – null and alternative hypotheses

Remember that in many instances it is easier to begin with H_a, the effect for which we hope to find evidence, and then to set up H_0 as the statement that the effect is not present. In each example, think carefully about whether H_a should be one–sided or two–sided.

a) The researcher hopes to show that on the average the mice complete the maze faster when there is a loud noise. What is H_a? What is H_0?

b) What does the teaching assistant hope to show? Use this to set up H_a and H_0.

c) Is the alternative one-sided or two-sided? Set up H_0 and H_a.

Exercise 6.39

KEY CONCEPTS – null and alternative hypotheses, carrying out a significance test about a mean

a) What do the researchers hope to show? This will be the alternative. Write down the null and alternative hypotheses.

b) The first step in carrying out the test is to compute the test statistic

$$z = \frac{\bar{x} - \mu_0}{\sigma / \sqrt{n}}$$

which measures how far the sample mean is from the hypothesized value μ_0. To find the numerical value of z, you need to determine μ_0, σ, and n from the problem, and then compute $\bar{x}$, the mean DRP score from the data given.

$z =$

Once the value of the test statistic z has been determined, the P-value can be computed. The P-value is the probability that the test statistic takes a value at least as extreme as the one observed. In the space provided, write this down as a probability in terms of Z, the standard normal, and then use Table A to evaluate this probability.

P-value $= P(Z \qquad) =$

If you are having trouble doing this directly from the meaning of the P-value, refer to the rules for computing P-values given in the box in the text and try to understand the rationale behind them.

Now try to interpret your result in plain language.

Exercise 6.41

KEY CONCEPTS – testing hypotheses at a fixed significance level

The value of the test statistic is given as $z = 2.42$. When carrying out the test at a fixed significance level, you can first compute the P-value and then reject the null hypothesis if the P-value is smaller than the significance level given. However, it is more direct to look up the critical value z^* in Table D and compare the value of the test statistic directly to the critical value.

a) The alternative is $\mu > 1.4$, so the P–value is $P(Z > 2.42)$, which represents the probability of a test statistic at least as extreme as the one observed. Compare the P-value to 0.05. What do you conclude?

Using Table D, we look up the critical value corresponding to the tail probability 0.05 (it does not need to be doubled since the test is one–sided) and find the critical value $z^* = 1.645$. We reject if the computed value of the test statistic exceeds this critical value. What do you conclude? When the significance level is fixed, it is easier to use Table D directly.

b) Follow the procedure in (a) but with significance level 1%.

Exercise 6.47

KEY CONCEPTS - calculating *P*-values

We need to compare the value $z = -1.37$ to the critical values in Table D. Since the test is two–sided, we use $|z| = 1.37$, and double the tail area. Its value is between the critical values $z^* = 1.282$ and $z^* = 1.645$. To what two significance levels do these critical values correspond? Remember that the test is two-sided. What can we say about the *P*-value?

Now use Table A to determine the *P*-value by computing the area to the left of -1.37. Is this area the *P*-value? Why or why not? If not, what do you need to do to find the *P*-value?

Exercise 6.49

KEY CONCEPTS - relationship between two–sided tests and confidence intervals

a) The form of the confidence interval is $\bar{x} \pm z^* \dfrac{\sigma}{\sqrt{n}}$, where $\bar{x}$ is the mean of the 24 weights, $\sigma = 4.5$ kg, $n = 24$ and z^* is the value on the normal curve with area C between $-z^*$ and z^*. Use Table D to find z^*, calculate $\bar{x}$ from the data, and use these to compute the interval.

Interval :

b) A level α two–sided significance test rejects a hypothesis $H_0: \mu = \mu_0$ when μ_0 falls outside a level $1 - \alpha$ confidence interval for μ. In this exercise, the test is at significance level $\alpha = 0.05$, which means we need to check if μ_0 is outside the $1 - 0.05 = 0.95$ or 95% confidence interval. The 95% confidence interval was computed in (a). What is the value of μ_0 and is it outside the 95% confidence interval? Do we reject at the 5% significance level?

c) What is μ_0 in this case? Can we reject $H_0: \mu = 63$ at the 5% significance level?

COMPLETE SOLUTIONS

Exercise 6.27

a) The researcher hopes to show that on the average the mice complete the maze faster when there is a loud noise: namely that $\mu < 18$. Thus the alternative hypothesis is H_a: $\mu < 18$ and the null hypothesis is H_0: $\mu = 18$, or it is equally correct to write the null hypothesis as H_0: $\mu \geq 18$.

b) The teaching assistant hopes to show that the students in classes he teaches will have a higher mean score than the class mean of 50. The hypotheses are H_0: $\mu = 50$ and H_a: $\mu > 50$.

c) The language department wants to see if there is evidence that the mean score for students who pass the credit examination is different from the mean score of students who take the course. The mean score of students who take the course is 24. The language department is interested in differences in either direction. The alternative is two-sided, and the hypotheses are H_0: $\mu = 24$ and H_a: $\mu \neq 24$.

Exercise 6.39

a) The researchers hope to show that their district mean exceeds the national mean of 32. The alternative is $\mu > 32$, so the hypotheses of interest to the researchers are H_0: $\mu = 32$ and H_a: $\mu > 32$.

b) The value of the test statistic is

$$z = \frac{\bar{x} - \mu_0}{\sigma / \sqrt{n}} = \frac{35.091 - 32}{11 / \sqrt{44}} = 1.86$$

Since the alternative is $\mu > 32$, the P-value is the chance of getting a value of $\bar{x}$ at least as large as 35.091 if the true mean were 32. In terms of the test statistic, this is equivalent to computing $P(Z \geq z) = P(Z \geq 1.86) = 1 - 0.9686 = 0.0314$.

There is evidence that the mean score of all third-graders in this district exceeds the national mean of 32.

Exercise 6.41

a) The value of the test statistic is given as $z = 2.42$. The P-value is $P(Z > 2.42) = 1 - 0.9922 = 0.0078$. Since the value is below 0.05, we reject at the 5% level of significance.

Using Table D, we look up the critical value corresponding to the tail probability 0.05 and find the critical value $z^* = 1.645$. A 5% level test rejects if

$z > 1.645$. Since 2.42 exceeds this value, we reject at the 5% level of significance

b) The P-value was computed in (a) to be 0.0078. Since $1\% = 0.01$ exceeds this value, we also reject at the 1% level of significance. This is the advantage of giving the P-value. It allows us to assess significance at any level.

Using Table D, we look up the critical value corresponding to the tail probability 0.01 and find the critical value $z^* = 2.326$. Since $z = 2.42$ exceeds the critical value, we reject at the 1% level of significance.

Exercise 6.47

From Table D, $z^* = 1.282$ corresponds to a significance level of $2 \times 0.10 = 0.20$, since the test is two-sided. The tabled value $z^* = 1.645$ corresponds to a significance level of $2 \times 0.05 = 0.10$. We would reject at $\alpha = 0.20$ since $|z| = 1.37 > 1.282$, but not at $\alpha = 0.10$ since $|z| = 1.37 < 1.645$. So the P-value lies between 0.10 and 0.20.

Using Table A, the area to the left of -1.37 is 0.0853 and the P-value is $2 \times 0.0853 = 0.1706$, since the test is two–sided.

Exercise 6.49

a) $\bar{x} = 61.79$ is the mean of the $n = 24$ weights, $\sigma = 4.5$ kg, and $z^* = 1.96$ from Table D, since $C = 0.95$ in this case. The interval is

$$61.79 \pm 1.96 \frac{4.5}{\sqrt{24}} = 61.79 \pm 1.80 = (59.99, 63.59).$$

b) Since H_0: $\mu = 61.3$, $\mu_0 = 61.3$, which doesn't fall outside the level $1 - \alpha = 0.95$ confidence interval for μ. So we fail to reject H_0 at significance level 5%.

c) Since H_0: $\mu = 63$, $\mu_0 = 63$, which doesn't fall outside the level $1 - \alpha = 0.95$ confidence interval for μ. So we fail to reject H_0 at significance level 5%.

SECTION 6.3

OVERVIEW

When describing the outcome of a hypothesis test it is more informative to give the P-value than just the reject or not decision at a particular significance level α. The traditional levels of 0.01, 0.05, and 0.10 are arbitrary and serve as rough guidelines.

When testing hypotheses with a very large sample, the *P*-value can be very small for effects that may not be of interest. Don't confuse small *P*-values with large or important effects. Plot the data to display the effect you are trying to show and also give a confidence interval which says something about the size of the effect.

Just because a test is not statistically significant doesn't imply that the null hypothesis is true. A test is not significant when it is based on a small sample size and has low power. Finally, if you run enough tests, you will invariably find statistical significance for one of them. Be careful in interpreting the results when testing many hypotheses on the same data.

GUIDED SOLUTIONS

Exercise 6.55

KEY CONCEPTS – statistical significance versus practical importance

In this problem, we see that the *P*-value associated with the outcome $\bar{x} = 478$ depends on the sample size. The probability of getting a value of $\bar{x}$ as large as 478 if the mean is 475 will become smaller as the sample size gets larger. Why? Since this probability is the *P*-value, we see that a small effect is more likely to be detected for larger sample sizes than for smaller sample sizes. But this doesn't necessarily make the effect interesting or important. A confidence interval, not the *P*-value, tells you something about the size of the effect.

a) Find the *P*-value by computing the z–statistic and the probability of exceeding it.

b) This is the same as (a) but with a larger sample size. The larger sample size makes the probability of getting a value of $\bar{x}$ as large as 478 smaller than it was in (a). Compute the *P*-value.

c) Find the *P*-value in this last case. It will be the smallest. Why?

Exercise 6.57

KEY CONCEPTS - significance levels vs. *P*-values

a) We reject the null hypothesis if the value of the z statistic exceeds 1.645 since the significance level is 0.05. Compute the test statistic when $\bar{x} = 491.4$. What do you conclude at $\alpha = 0.05$?

b) Compute the test statistic when $\bar{x} = 491.5$. What do you conclude at $\alpha = 0.05$.

Exercise 6.61

KEY CONCEPTS - multiple tests, the *Bonferroni* procedure

If you perform k tests and want protection at level α, use α/k as your cutoff for statistical significance. In our case, k = 12 and $\alpha = 0.05$ so each test is conducted at the 0.05/12 = 0.00417 significance level. Which *P*-values given lead to rejection at the 0.00417 level?

COMPLETE SOLUTIONS

Exercise 6.55

a) The test statistic is $z = \dfrac{\bar{x} - \mu_0}{s/\sqrt{n}} = \dfrac{478 - 475}{100/\sqrt{100}} = 0.3$, and
P-value = $P(Z > 0.3) = 1 - 0.6179 = 0.3821$ since the alternative is one sided.

b) The test statistic is $z = \dfrac{\bar{x} - \mu_0}{s/\sqrt{n}} = \dfrac{478 - 475}{100/\sqrt{1000}} = 0.95$, and
P-value = $P(Z > 0.95) = 1 - 0.8289 = 0.1711$.

c) The test statistic is $z = \dfrac{\bar{x} - \mu_0}{s/\sqrt{n}} = \dfrac{478 - 475}{100/\sqrt{10000}} = 3$, and
P-value = $P(Z > 3) = 1 - 0.9987 = 0.0013$.

Exercise 6.57

In this problem we see that the *P*-value is more informative than just stating whether we accept or reject at $\alpha = 0.05$. In both (a) and (b) the evidence against the null hypothesis is almost the same, yet at the 5% level of significance we reject in (b) but not in (a). Don't attach too much importance to the 5% level of significance. Use the *P*-value and think about the problem in the context of the field and in terms of other data that may have been collected.

a) $z = \dfrac{\bar{x} - \mu_0}{s/\sqrt{n}} = \dfrac{491.4 - 475}{100/\sqrt{100}} = 1.64$, so we don't reject at $\alpha = 0.05$. The *P*-value is 0.0505.

b) $z = \dfrac{\bar{x} - \mu_0}{s/\sqrt{n}} = \dfrac{491.5 - 475}{100/\sqrt{100}} = 1.65$, so we reject at $\alpha = 0.05$. The *P*-value is 0.0495.

Exercise 6.61

If the *P*-value is less than $\alpha/k = 0.05/12 = 0.00417$, then we reject using the *Bonferroni* procedure. In this case we reject the null hypothesis for three of the tests - the ones with *P*-values of 0.001, 0.004, and 0.002, since these *P*-values are smaller than 0.00417.

SECTION 6.4

OVERVIEW

The **power** of a significance test is always calculated at a specific alternative hypothesis and is the probability that the test will reject H_0 when that alternative is true. This calculation requires knowledge of the sampling distribution under the specific alternative hypothesis of the test statistic used. Power is usually interpreted as the ability of a test to detect an alternative hypothesis or as the sensitivity of a test to an alternative hypothesis. The power of a test can be increased by increasing the sample size when the significance level remains fixed.

To compute the power of a significance test about a mean of a normal population, we need to

- state H_0, H_a (the particular alternative we want to detect), and the significance level α,

- find the values of $\bar{x}$ that will lead us to reject H_0,

- calculate the probability of observing these values of $\bar{x}$ when the alternative is true.

Decision theory regards statistical inference as giving rules for making decisions in the presence of uncertainty. From the decision theory point of view, H_0 and H_a are just two statements of equal status that we must decide between. Decision analysis chooses a rule for deciding between H_0 and H_a on the basis of the probabilities of the two types of errors that we can make. A **Type I error** occurs if H_0 is rejected when it is in fact true. A **Type II error** occurs if H_0 is accepted when in fact H_a is true.

There is a clear relation between α level significance tests and testing from the decision theory point of view. α is the probability of a Type I error, and the power of the test against a specific alternative is 1 minus the probability of a Type II error for that alternative.

GUIDED SOLUTIONS

Exercise 6.63

KEY CONCEPTS - power of a significance test

To compute the power of a significance test about a mean, we need to

(i) State H_0, H_a (the particular alternative we want to detect) and the significance level α.

(ii) Find the values of $\bar{x}$ that will lead us to reject H_0.

(iii) Calculate the probability of observing these values of $\bar{x}$ when the alternative is true.

Following these steps, we notice:

(i) In this problem, the hypotheses are

$$H_0: \mu = 450 \text{ vs. } H_a: \mu > 450$$

The particular alternative we want to detect is $\mu = 460$. The significance level is $\alpha = 0.01$.

(ii) The values of $\bar{x}$ that will lead us to reject H_0 are indicated in the problem and are those for which

$$z = \frac{\bar{x} - 450}{100 / \sqrt{500}} \geq 2.326$$

since $\sigma = 100$ and the sample size $n = 500$. Solve the above inequality for one of the form $\bar{x} \geq$??.

(iii) The probability of observing these values of $\bar{x}$ when the alternative is true is

$$P(\bar{x} \geq 460.4 \text{ when } \mu = 460) = P(\frac{\bar{x} - \mu}{\sigma / \sqrt{n}} \geq \frac{460.4 - 460}{100 / \sqrt{500}})$$

Now complete the calculation of this probability using Table A. The result will be the desired power.

Exercise 6.67

KEY CONCEPTS – Type I and Type II error probabilities

a) From the information given in the problem we are told that our population is normal; our sample size is $n = 9$, $\sigma = 1$; we are testing $H_0: \mu = 0$ vs. $H_a: \mu > 0$; and we reject H_0 if $\bar{x} > 0$. Thus

$$P(\text{Type I error}) = P(\text{our test rejects } H_0 \text{ when in fact } \mu = 0)$$

$$= P(\bar{x} > 0 \text{ when in fact } \mu = 0)$$

Now compute this probability.

b) In this part, we want (since we accept H_0 if $\bar{x} \leq 0$)

$$P(\text{Type II error}) = P(\text{our test accepts } H_0 \text{ when in fact } \mu = 0.3)$$

$$= P(\bar{x} \leq 0 \text{ when in fact } \mu = 0.3)$$

Now compute this probability.

c) See if you can do this on your own. Repeat the argument in (b) with $\mu = 1$ in place of $\mu = 0.3$.

COMPLETE SOLUTIONS

Exercise 6.63

We follow the three steps indicated in the guided solutions.

(i) This step was completely discussed in the guided solutions.

(ii) In terms of $\bar{x}$, after solving the inequality given in the guided solutions, we reject H_0 if $\bar{x} \geq (2.326 \times 100 / \sqrt{500}) + 450 = 460.4$.

(iii) We find

$$P(\bar{x} \geq 460.4 \text{ when } \mu = 460) = P(\frac{\bar{x} - \mu}{\sigma / \sqrt{n}} \geq \frac{460.4 - 460}{100 / \sqrt{500}})$$

$$= P(Z \geq 0.09) = 0.4641$$

where we have used Table A to compute $P(Z \geq 0.09)$. This is the desired power.

Exercise 6.67

a) We have

$$P(\text{Type I error}) = P(\text{our test rejects } H_0 \text{ when in fact } \mu = 0)$$

$$= P(\bar{x} > 0 \text{ when in fact } \mu = 0)$$

$$= P(\frac{\bar{x} - \mu}{\sigma / \sqrt{n}} > \frac{0 - 0}{1 / \sqrt{9}})$$

$$= P(Z > 0)$$

which we know is 0.5.

b) We have

$P($Type II error$)\quad = P($our test accepts H_0 when in fact $\mu = 0.3)$

$$= P(\bar{x} \leq 0 \text{ when in fact } \mu = 0.3)$$

$$= P(\frac{\bar{x} - \mu}{\sigma / \sqrt{n}} \leq \frac{0 - 0.3}{1 / \sqrt{9}})$$

$$= P(Z \leq -0.9)$$

$$= 0.1841$$

c) We have

$P($Type II error$)\quad = P($our test accepts H_0 when in fact $\mu = 1)$

$$= P(\bar{x} \leq 0 \text{ when in fact } \mu = 1)$$

$$= P(\frac{\bar{x} - \mu}{\sigma / \sqrt{n}} \leq \frac{0 - 1}{1 / \sqrt{9}})$$

$$= P(Z \leq -3)$$

$$= 0.0013$$

CHAPTER 7

INFERENCE FOR DISTRIBUTIONS

SECTION 7.1

OVERVIEW

Confidence intervals and significance tests for the mean μ of a normal population are based on the sample mean $\bar{x}$ of an SRS. When the sample size n is large, the central limit theorem suggests that these procedures are approximately correct for other population distributions. In Chapter 6, we considered the (unrealistic) situation in which we knew the population standard deviation σ. In this section, we consider the more realistic case where σ is not known and we must estimate σ from our SRS by the sample standard deviation s. In Chapter 6 we used the **one-sample z -statistic**

$$z = \frac{\bar{x} - \mu}{\sigma / \sqrt{n}}$$

which has the $N(0,1)$ distribution. Replacing σ by s, we now use the **one-sample t -statistic**

$$t = \frac{\bar{x} - \mu}{s / \sqrt{n}}$$

which has the *t* **distribution** with *n* - 1 **degrees of freedom**.

For every positive value of *k*., there is a *t* distribution, with *k* degrees of freedom, denoted *t(k)*. All are symmetric, bell-shaped distributions, similar in shape to normal distributions but with greater spread. As *k* increases, *t(k)* approaches the $N(0,1)$ distribution.

A level *C* **confidence interval for the mean** μ of a normal population when σ is unknown is

$$\bar{x} \pm t^* \frac{s}{\sqrt{n}}$$

where t^* is the upper $(1-C)/2$ critical value of the $t(n-1)$ distribution whose value can be found in Table D of the text or from statistical software. The quantity $t^* \dfrac{s}{\sqrt{n}}$ is the **margin of error**.

Significance tests of H_0: $\mu = \mu_0$ are based on the one-sample *t* statistic. Such tests are often referred to as **one-sample *t* tests**. *P*-values or fixed significance levels are computed from the $t(n-1)$ distribution using Table D or, more commonly in practice, using statistical software.

The power of the *t* test is calculated like that of the *z* test as described in Chapter 6, using an approximate value (perhaps based on past experience or a pilot study) for both σ and *s*.

One application of these one-sample *t* procedures is to the analysis of data from **matched pairs** studies. We compute the differences between the two values of a matched pair (often before and after measurements on the same unit) to produce a single sample value. The sample mean and standard deviation of these differences are computed. Depending on whether we are interested in a confidence interval or a test of significance concerning the difference in the population means of matched pairs, we either use the one-sample confidence interval or the one-sample significance test based on the *t* statistic.

For larger sample sizes, the *t* procedures are fairly **robust** against nonnormal populations. As a rule of thumb, *t* procedures are useful for nonnormal data when $n \geq 15$ unless the data show outliers or strong skewness, and for samples of size $n \geq 40$ *t* procedures can be used for even clearly skewed distributions. Data consisting of small samples from skewed populations can sometimes be analyzed by first applying a **transformation** (such as logarithms) to the data to obtain an approximately normally distributed variable. The *t* procedures can then be applied to the transformed data. When transformations are used, it is a good idea to examine stemplots, histograms, or normal quantile plots of the transformed data to verify that the transformed data appear to be approximately normally distributed.

Another procedure that can be used with smaller samples from a non-normal population is the **sign test**. The sign test is most useful for testing for "no treatment effect" in matched pairs studies. As with the matched pairs t test, one computes the differences of the two values in each matched pair. Pairs with difference 0 are ignored, and the number of trials n is the count of the remaining pairs. The test statistic is the count X of pairs with a positive difference. P-values for X are based on the binomial $B(n, 1/2)$ distribution. The sign test is less powerful than the t test in cases where the use of the t test is justified.

GUIDED SOLUTIONS

Exercise 7.1

KEY CONCEPTS – critical values of the $t(k)$ distribution

a) For confidence intervals, the critical value t^* is the upper $(1 - C)/2$ critical value of the $t(n - 1)$ distribution. We must identify the confidence level C and the sample size n. Since we are interested in a 90% confidence interval, C is 0.90. Since we take 12 observations, $n = 12$. Thus t^* is the upper $(1 - 0.90)/2 = 0.05$ critical value of the $t(12 - 1) = t(11)$ distribution. Now turn to Table D and look under the column labeled 0.05 (the desired tail probability) and in the row labeled 11 (the degrees of freedom). What value do you find in the table?

b) This is just like (a). Identify C and n here and proceed as in (a).

c) Again, this is just like (a). See if you can do this on your own.

Exercise 7.11

KEY CONCEPTS - confidence intervals based on the one–sample t statistic, checking assumptions, robustness of t procedures

a) A stemplot and normal quantile plot are most easily constructed using statistical software. Features to look for are outliers and symmetry versus skewness. You may wish to refer to Section 1.3 to refresh your memory about how to interpret normal quantile plots. You can do the stemplot by hand below. What are the important features of the stemplot and normal probability plot?

b) To compute a 95% confidence interval, we use the formula

$$\bar{x} \pm t^* \frac{s}{\sqrt{n}}$$

where t^* is the upper $(1 - C)/2$ critical value of the $t(n - 1)$ distribution. Here C is the confidence level of 0.95, and n is the sample size of 50. Thus in this exercise, t^* is the upper $(1 - 0.95)/2 = 0.025$ critical value of the $t(50 - 1) = t(49)$ distribution. Find this critical value in Table D.

$t^* =$

Next compute $\bar{x}$ and s, preferably using statistical software or a calculator.

$\bar{x} =$ $s =$

Finally, substitute all these values into the formula to complete the computation of the 95% confidence interval.

$$\bar{x} \pm t^* \frac{s}{\sqrt{n}} =$$

Note that a much easier approach would be to use statistical software to compute the 95% confidence interval directly. Consult the manual for your software for instructions.

c) Review Section 7.1 regarding the robustness of t procedures. Are the conditions for robustness satisfied here?

Exercise 7.25

KEY CONCEPTS - one sample t tests

We are interested in whether or not the patient's mean level μ falls above 4.8. What effect do you wish to show? Use this to set up the null and alternative hypotheses

H_0:
H_a:

Make sure you understand why we use a one-sided alternative here.

The sample mean and standard deviation for the data given Exercise 7.23 are easily calculated to be

$\bar{x} = 5.367$ $s = 0.665$

and the standard error is

$$\frac{s}{\sqrt{n}} = \frac{0.665}{\sqrt{6}} = 0.271.$$

Use this information to calculate the one sample t statistic given below.

$$t = \frac{\bar{x} - \mu}{s/\sqrt{n}} =$$

Based on the value of this t statistic, between which levels from Table D does the P-value lie? What conclusions do you draw?

Exercise 7.33

KEY CONCEPTS - matched pairs experiments, one-sample t tests

a) This is a matched pairs experiment. The matched pair of observations are the right and left thread measurements on each subject. To avoid confounding with time, we would probably want subjects to use both knobs in the same session. We would also want to randomize which knob the subject uses first. How might you do this randomization?

b) The project hopes to show that right–handed people find right–hand threads easier to use than left–hand threads. In terms of the mean μ for the population of differences

$$\text{(left thread time)} - \text{(right thread time)}$$

what do we wish to show? This would be the alternative. What are H_0 and H_a (in terms of μ)?

c) For data from a matched pairs study, we compute the differences between the two values of a matched pair to produce a single sample value. These differences are given below for our data.

Right thread	Left thread	Difference = Left – Right
113	137	24
105	105	0
130	133	3
101	108	7
138	115	−23
118	170	52
87	103	16
116	145	29
75	78	3
96	107	11
122	84	−38
103	148	45
116	147	31
107	87	−20
118	166	48
103	146	43
111	123	12
104	135	31
111	112	1
89	93	4
78	76	−2
100	116	16
89	78	−11
85	101	16
88	123	35

The sample mean and standard deviation of these differences are now computed.

$$\bar{x} = \qquad\qquad s =$$

We then use the one sample significance test based on the t statistic. What value of μ should be used

$$t = \frac{\bar{x} - \mu}{s/\sqrt{n}} =$$

From the value of the t statistic and Table D (or using statistical software), the P-value can be computed.

P-value =

What conclusion do you draw?

Note: This problem is most easily done directly using statistical software. The software will compute the differences, the t statistic and the P-value for you. Consult your users manual to see how to do one-sample t tests.

Exercise 7.39

KEY CONCEPTS - appropriateness of statistical procedures

Statistical procedures use information in a sample to make inferences about parameters of a population. Ask yourself, what is the population and the parameter of interest in this exercise? What does our data allow us to say about this parameter?

Exercise 7.41

KEY CONCEPTS - power of the one-sample t test

a) You may wish to refer to Section 6.4 to review power. We wish to determine the power of the test against the alternative $\mu = 0.5$ when the significance level is 0.05 when $n = 10$. We use 0.83 as an estimate of both the population standard deviation σ and s in future samples. The t test with 10 observations rejects H_0: $\mu = 0$ (this is the null hypothesis that the mean difference in yields is 0) if the t statistic

$$t = \frac{\bar{x} - 0}{s/\sqrt{10}}$$

exceeds the upper 5% point of $t(9)$, which is 1.8331 (obtained using statistical software, but you can use Table D to get $t(9)$ also). Using $s = 0.83$, the event that the test rejects H_0 is

$$t = \frac{\bar{x}}{0.83/\sqrt{10}} \geq 1.8331$$

or

$$\bar{x} \geq 1.8331 \frac{0.83}{\sqrt{10}} = 0.481$$

The power is the probability that $\bar{x} \geq 0.481$ when $\mu = 0.5$. Taking $\sigma = 0.83$, this probability is found by standardizing $\bar{x}$,

$$P(\bar{x} \geq 0.481 \text{ when } \mu = 0.5) \quad = P(\frac{\bar{x} - 0.5}{0.83/\sqrt{10}} \geq \frac{0.481 - 0.5}{0.83/\sqrt{10}})$$

$$= P(Z \geq -0.072) = 1 - 0.4713 = 0.5287$$

Thus, the power is 0.5287.

b) Try this one on your own. Use the same argument as in (a) but now with $n = 25$ rather than $n = 10$ wherever the value of n appears. All other features of the problem are the same.

Exercise 7.43

KEY CONCEPTS - the sign test

a) In Exercise 7.33, we investigated whether the mean μ for the population of differences

(time to complete task with left–hand thread)
 – (time to complete task with right–hand thread)

for matched pairs was greater than 0. Recall that this led to the hypotheses

H_0: $\mu = 0$ and H_a: $\mu > 0$

Try to formulate analogous hypotheses in terms of

  • The median for the population of these differences

  • The probability p of completing the task faster with the right-hand thread

What do the statements H_0: $\mu = 0$ and H_a: $\mu > 0$ imply about the median for the population of differences? Remember, the median and mean both attempt to measure the center of a population distribution.

What do the statements H_0: $\mu = 0$ and H_a: $\mu > 0$ imply about the probability p of completing the task faster with the *right–hand* thread?

b) Recall that for the sign test, pairs with difference 0 are ignored and the number of trials n is the count of the remaining pairs. The test statistic is the count X of pairs with a positive difference, since this occurs if the right–hand thread takes less time than the left–hand thread. P-values for X are based on the binomial $B(n, 1/2)$ distribution.

Refer to the guided solution for Exercise 7.33 where the 25 differences are given. How many pairs with difference 0 are there in the data? What, therefore, is n?

What is the observed number of pairs with a positive difference in our data?

In terms of the probability p of completing the task faster with the right–hand thread, we wish to test the hypotheses H_0: $p = 1/2$ and H_a: $p > 1/2$. This implies that the P-value is

$P(X \geq$ observed number of pairs with a positive difference in our data)

Compute this probability. Note that although X has the binomial $B(n, 1/2)$ distribution, since n is larger than 20, we can use the normal approximation to the binomial distribution to compute the P-value.

COMPLETE SOLUTIONS

Exercise 7.1

a) The value of t^* that you should find in Table D is 1.796.

b) Since we are interested in a 95% confidence interval, C is 0.95. Since we take an SRS of 30 observations, $n = 30$. The desired critical value t^* is the upper $(1 - 0.95)/2 = 0.025$ critical value of the $t(30 - 1) = t(29)$ distribution. Now turn to Table D and look in the column labeled 0.025 and the row labeled 29. You should find

$$t^* = 2.045$$

c) Since we are interested in an 80% confidence interval, C is 0.80. Since we take a sample of size 18, $n = 18$. Thus t^* is the upper $(1 - 0.80)/2 = 0.10$ critical value of the $t(18 - 1) = t(17)$ distribution. Now turn to Table D and look in the column labeled 0.10 and the row labeled 17. You should find

$$t^* = 1.333$$

Exercise 7.11

a) A stemplot of the data is given below. We have used split stems because of the large number of observations between 10 and 19.

```
1 | 01233344
1 | 5566667778999999
2 | 00124444
2 | 5555566667
3 | 244
3 | 5
4 | 1
4 | 8
5 |
5 |
6 | 3
6 |
7 |
7 | 9
```

A normal quantile plot of these data is

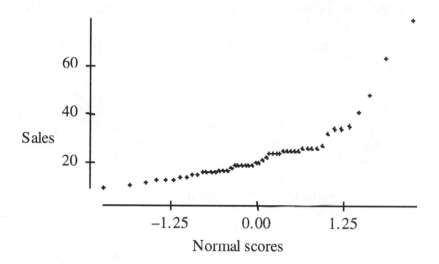

Both plots indicate that the data are right-skewed.

b) The desired quantities are

$$t^* = 2.0096 \qquad \bar{x} = 23.56 \qquad s = 12.52$$

$$\bar{x} \pm t^* \frac{s}{\sqrt{n}} = 23.56 \pm (2.0096)\frac{12.52}{\sqrt{50}} = 23.56 \pm 3.56$$

c) As a rule of thumb, t procedures are useful for non-normal data when $n \geq 15$, unless the data show outliers or strong skewness, and for samples of size $n \geq 40$, t procedures can be used even for clearly skewed distributions. In this exercise $n = 50$. Thus, even though there is clear skewness in the distribution, it is reasonably safe to use t procedures here.

Exercise 7.25

The hypotheses we wish to test are

$H_0: \mu = 4.8$

$H_a: \mu > 4.8$

We compute

$$t = \frac{\bar{x} - \mu}{s/\sqrt{n}} = \frac{5.367 - 4.8}{0.271} = 2.092$$

According to the row corresponding to $n - 1 = 5$ in Table D, the P-value lies between 0.05 and 0.025. Using statistical software, we find that the P-value is 0.0453. We conclude that there is good evidence that the patient's mean level μ in fact falls above 4.8.

Exercise 7.33

a) The randomization might be carried out simply by flipping a fair coin. If the coin comes up heads, use the right–hand threaded knob first. If the coin comes up tails, use the left–hand threaded knob first.

b) In terms of μ, the mean of the population of differences, (left thread time) – (right thread time), we wish to test if the times for the left-threaded knobs are longer than those for the right threaded knobs; i.e.,

$$H_0: \mu = 0 \text{ vs. } H_a: \mu > 0$$

c) For the 25 differences, we compute

$$\bar{x} = 13.32 \qquad\qquad s = 22.94$$

We then use the one-sample significance test based on the t-statistic.

$$t = \frac{\bar{x} - \mu}{s/\sqrt{n}} = \frac{13.32 - 0}{22.94/\sqrt{25}} = 2.903$$

From the value of the t-statistic and Table D, the P-value is between 0.0025 and 0.005. Using statistical software, the P-value is computed as P-value $= 0.0039$.

We conclude that there is strong evidence that the time for left-hand threads is greater than that for right-hand threads, on average.

Exercise 7.39

The population is all 50 states. The parameter of interest is mean percentage of over-65 residents in this population. In this case, our data percentage of over-65 residents for the entire population. Thus, we can compute the desired parameter exactly. No inference is needed.

Exercise 7.41

a) A complete solution is given in the guided solution.

b) The *t* test with 25 observations rejects H_0: $\mu = 0$ (this is the null hypothesis that the mean difference in yields is 0) if the *t*-statistic

$$t = \frac{\bar{x} - 0}{s/\sqrt{25}}$$

exceeds the upper 5% point of $t(24)$, which is 1.711 (obtained using statistical software). Using $s = 0.83$, the event that the test rejects H_0 is

$$t = \frac{\bar{x}}{0.83/\sqrt{25}} \geq 1.711$$

or

$$\bar{x} \geq 1.711 \frac{0.83}{\sqrt{25}} = 0.284$$

The power is the probability that $\bar{x} \geq 0.284$ when $\mu = 0.5$. Taking $\sigma = 0.83$, this probability is found by standardizing $\bar{x}$,

$$P(\bar{x} \geq 0.284 \text{ when } \mu = 0.5) \quad = P(\frac{\bar{x} - 0.5}{0.83/\sqrt{25}} \geq \frac{0.284 - 0.5}{0.83/\sqrt{25}})$$

$$= P(Z \geq -1.301) = 1 - 0.0966 = 0.9034$$

Thus, the power is 0.9034.

Exercise 7.43

a) In terms of the median for the population of differences

> (time to complete task with left–hand thread)
> – (time to complete task with right–hand thread)

the hypotheses are

H_0: population median $= 0$ and H_a: population median > 0

since these hypotheses test whether there is a tendency for the times for the left–hand thread to be longer than the times for the right–hand thread.

In terms of the probability p of completing the task faster with the *right–hand* thread, the hypotheses would be

$$H_0: p = 1/2 \text{ and } H_a: p > 1/2$$

since these hypotheses test whether we are more likely to complete the task faster with the right–hand thread.

b) Since there is only one pair with difference 0 in the data, $n = 24$. The number of pairs with a positive difference in our data is 19. Thus, the P-value is

$$P(X \geq 19) = 1 - P(X \leq 18)$$

where X has the binomial $B(24, 1/2)$ distribution. We use the normal approximation to the binomial to compute this probability. X has mean

$$\mu = np = (24)(1/2) = 12$$

and standard deviation

$$\sigma = \sqrt{np(1-p)} = \sqrt{(24)(1/2)(1/2)} = \sqrt{6} = 2.45$$

The z-score of 18 is thus

$$z = \frac{18 - \mu}{\sigma} = \frac{18 - 12}{2.45} = 2.45$$

and so our P-value is

$$
\begin{aligned}
P(X \geq 19) \quad &= 1 - P(X \leq 18) \\
&= 1 - P(\text{standard normal random variable} \leq 2.45) \\
&= 1 - 0.9929 \\
&= 0.0071
\end{aligned}
$$

SECTION 7.2

OVERVIEW

One of the most commonly used significance tests is the **comparison of two population means,** μ_1 and μ_2. In this setting we have two distinct, independent SRS's from two populations or two treatments on two samples. The procedures are based on the difference $\bar{x}_1 - \bar{x}_2$. When the populations are not normal, the results obtained using the methods of this section are approximately correct due to the central limit theorem.

Tests and confidence intervals for the difference in the population means, $\mu_1 - \mu_2$, are based on the **two-sample t statistic**,

$$t = \frac{(\bar{x}_1 - \bar{x}_2) - (\mu_1 - \mu_2)}{\sqrt{\dfrac{s_1^2}{n_1} + \dfrac{s_2^2}{n_2}}}$$

Despite the name, this test statistic does *not* have an exact t distribution. However there are good approximations to its distribution which allow us to carry out valid significance tests. Conservative procedures use the $t(k)$ distribution as an approximation where the degrees of freedom k is taken to be the smaller of $n_1 - 1$ and $n_2 - 1$. More accurate procedures use the data to estimate the degrees of freedom k. This is the procedure that is followed by most statistical software.

To carry out a significance test for $H_0 : \mu_1 = \mu_2$, use the two-sample t statistic

$$t = \frac{(\bar{x}_1 - \bar{x}_2)}{\sqrt{\dfrac{s_1^2}{n_1} + \dfrac{s_2^2}{n_2}}}$$

The P-value is found by using the approximate distribution $t(k)$, where k is estimated from the data when using statistical software, or can be taken to be the smaller of $n_1 - 1$ and $n_2 - 1$ for a conservative procedure.

An approximate confidence C level **confidence interval** for $\mu_1 - \mu_2$ is given by

$$(\bar{x}_1 - \bar{x}_2) \pm t^* \sqrt{\frac{s_1^2}{n_1} + \frac{s_2^2}{n_2}}$$

where t^* is the upper $(1 - C)/2$ critical value for $t(k)$, where k is estimated from the data when using statistical software, or can be taken to be the smaller of $n_1 - 1$ and $n_2 - 1$ for a conservative procedure. The procedures are most robust to failures in the assumptions when the sample sizes are equal.

The **pooled two-sample t procedures** are used when we can safely assume that the two populations have equal variances. The modifications in the procedure are the use of the pooled estimator of the common unknown variance

$$s_p^2 = \frac{(n_1 - 1)s_1^2 + (n_2 - 1)s_2^2}{n_1 + n_2 - 2}$$

and critical values obtained from the $t (n_1 + n_2 -2)$ distribution.

GUIDED SOLUTIONS

Exercise 7.55

KEY CONCEPTS - tests using the two-sample t, generalizations beyond the data

a) When means and standard deviations for the two samples are given, it is relatively easy to compute the value of the two-sample t statistic. With raw data as in this exercise, the details of the computations are best left to statistical software. However, although you may let the computer do the tedious computations, it is still important to know what all the numbers in the output mean and how to interpret the results. The output below is reproduced from MINITAB. The output is fairly standard between the different software packages.

Two Sample T-Test and Confidence Interval

```
Twosample T for Low vs High
          N       Mean      StDev    SE Mean
Low      14      4.640      0.690      0.18
High     14      6.429      0.430      0.12


T-Test mu Low = mu High (vs not =): T= -8.23 P=0.0000 DF = 21.77
```

The output begins with summary information for the two samples. For the Low group, $n_1 = 14$, $\bar{x}_1 = 4.640$, $s_1 = 0.690$ and the SE Mean is $s_1 / \sqrt{n_1} = 0.18$. For the High group, $n_2 = 14$, $\bar{x}_2 = 6.429$ $s_2 = 0.430$ and the SE Mean is $s_2 / \sqrt{n_2} = 0.12$. The next line gives the hypotheses, test statistic and P-value:

mu Low = mu High corresponds to H_0: $\mu_{Low} = \mu_{High}$, and

(vs not =) corresponds to H_a: $\mu_{Low} \neq \mu_{High}$.

Next is $t = \dfrac{\bar{x}_1 - \bar{x}_2}{\sqrt{\dfrac{s_1^2}{n_1} + \dfrac{s_2^2}{n_2}}} = \dfrac{4.640 - 6.429}{\sqrt{\dfrac{(0.690)^2}{14} + \dfrac{(0.430)^2}{14}}} = -8.23$ and the P-value =

0.0000. The degrees of freedom is found using the formula

$$\frac{\left(\dfrac{s_1^2}{n_1}+\dfrac{s_2^2}{n_2}\right)^2}{\dfrac{1}{n_1-1}\left(\dfrac{s_1^2}{n_1}\right)^2+\dfrac{1}{n_2-1}\left(\dfrac{s_2^2}{n_2}\right)^2}=\frac{\left(\dfrac{0.690^2}{14}+\dfrac{0.430^2}{14}\right)^2}{\dfrac{1}{14-1}\left(\dfrac{0.690^2}{14}\right)^2+\dfrac{1}{14-1}\left(\dfrac{0.430^2}{14}\right)^2}=21.77$$

You should make sure you understand how all the numbers in the computer output are obtained. Now use the information in the output to answer the questions: Is the difference in mean ego strength significant at the 5% level? At the 1% level? Why?

b) Are these groups random samples from low-fitness and high-fitness groups of middle-aged men? Where did the individuals in the samples come from? How might this affect generalizing to the population of all middle-aged men?

Exercise 7.71

KEY CONCEPTS - tests and confidence intervals using the two-sample t, relationship between two–sided tests and confidence intervals

The summary statistics for the comparison of child and adult VOT scores are reproduced below.

Group	n	$\bar{x}$	s
Children	10	–3.67	33.89
Adults	20	–23.17	50.74

a) Recall that the standard error of the mean is $s/\sqrt{n}$, where s is the sample standard deviation. Use this to compute the standard error of the sample mean VOT for the adults.

The standard error for the difference in two means $\bar{x}_1 - \bar{x}_2$ is

$$\sqrt{\frac{s_1^2}{n_1}+\frac{s_2^2}{n_2}}$$

Apply this to compute the standard error for the difference between the mean VOT for children and adults. Don't forget to square both standard deviations when evaluating the formula for the standard error.

b) The researchers were interested in a difference in either direction. Use this to set up the null and alternative hypotheses. Find the numerical value of the two-sample t statistic using the formula below. Remember that you found the standard error of the difference between the means in (a) which corresponds to the denominator of the t.

$$t = \frac{\bar{x}_1 - \bar{x}_2}{\sqrt{\dfrac{s_1^2}{n_1} + \dfrac{s_2^2}{n_2}}} =$$

Compute the P-value using a t distribution with the smaller of $n_1 - 1 = 9$ and $n_2 - 1 = 19$ degrees of freedom, and comparing the computed value of t to the critical values given in Table D. Remember to double the tail probabilities since the test is two-sided. State your conclusions.

c) The formula for the 95% confidence interval is $\bar{x}_1 - \bar{x}_2 \pm t * \sqrt{\dfrac{s_1^2}{n_1} + \dfrac{s_2^2}{n_2}}$, where $t*$ is the upper $(1-C)/2 = 0.025$ critical value for the t distribution with degrees of freedom equal to the smaller of $n_1 - 1 = 9$ and $n_2 - 1 = 19$. First find $t*$ from Table D, and complete the computations for the confidence interval. Remember the relationship between two-sided tests and confidence intervals. We reject the null hypothesis $H_0: \mu_1 - \mu_2 = 0$ at significance level α when the $1 - \alpha$ confidence interval for $\mu_1 - \mu_2$ doesn't contain the value 0.

Exercise 7.75

KEY CONCEPTS - pooled t test

The summary statistics for the comparison of male and female scores on the Chapin Social Insight Test are reproduced below.

Group	Sex	n	$\bar{x}$	s
1	M	133	25.34	5.05
2	F	162	24.94	5.44

The hypotheses are $H_0 : \mu_1 = \mu_2$ and $H_a : \mu_1 \neq \mu_2$. The computations for the pooled two–sample t test require the pooled estimator of σ^2, which combines the estimates of σ^2 from the two samples. The formula is

$$s_p^2 = \frac{(n_1 - 1)s_1^2 + (n_2 - 1)s_2^2}{n_1 + n_2 - 2} =$$

where you need to remember to square the standard deviations. The next step is to compute the value of the test statistic. The formula is

$$t = \frac{\bar{x}_1 - \bar{x}_2}{s_p \sqrt{\dfrac{1}{n_1} + \dfrac{1}{n_2}}} =$$

where you need to remember to use s_p, the square root of s_p^2, in the denominator. The P-value is found by comparing the computed value of t to critical values for the $t(n_1 + n_2 - 2) = t(293)$ distribution and then doubling the tail area since the alternative is two–sided. Use Table D to arrive at your answer.

COMPLETE SOLUTIONS

Exercise 7.55

a) If we know the P-value, we can assess the significance at any level by comparing the P-value to the significance level. Since the P-value is the smallest α level at which we can reject H_0, if the P-value is smaller than the significance level given, we reject H_0; otherwise, we do not reject. Since the P-value in the output is given as 0.0000, it is smaller than both 5% and 1%. So we reject H_0 at both these significance levels.

b) The individuals in the study are not random samples from low-fitness and high-fitness groups of middle-aged men. There are two ways in which they might be systematically different. The first is that all subjects in the study are college faculty. The "ego strength" personality factor for low-fitness and high-fitness middle-aged college faculty members might differ from the general population, and so might the mean difference between the groups. In addition, we don't have random samples of college faculty members. We are using volunteers in the study, and they might differ from both college faculty and the general population. It is hard to say in which direction this might bias the results, but the possibility of bias is definitely present.

Exercise 7.71

Group	n	$\bar{x}$	s
Children	10	-3.67	33.89
Adults	20	-23.17	50.74

a) The standard error of the sample mean VOT for the adults is

$$s / \sqrt{n} = 50.74 / \sqrt{20} = 11.346$$

The standard error for the difference $\bar{x}_1 - \bar{x}_2$ between the mean VOT for children and adults is

$$\sqrt{\frac{s_1^2}{n_1} + \frac{s_2^2}{n_2}} = \sqrt{\frac{33.89^2}{10} + \frac{50.74^2}{20}} = 15.607$$

b) The hypotheses are $H_0: \mu_1 = \mu_2$ versus $H_a: \mu_1 \neq \mu_2$, since the researchers were interested in a difference in either direction. The numerical value of the two-sample t statistic is given below where we have used the fact that the denominator of the t was computed in (a).

$$t = \frac{\bar{x}_1 - \bar{x}_2}{\sqrt{\frac{s_1^2}{n_1} + \frac{s_2^2}{n_2}}} = \frac{-3.67 - (-23.17)}{15.607} = 1.25$$

The smaller of $n_1 - 1 = 9$ and $n_2 - 1 = 19$ is 9 so we can refer the computed value of t to a $t(9)$ in Table D. The value 1.25 is between the critical values corresponding to upper tail probablities of 0.10 and 0.15. This gives a P-value between 0.20 and 0.30 We need to double upper tail probabilities from the table because the test is two-sided (The exact P-value for a t with 9 degrees of freedom using statistical software is 0.2428) The data give no evidence of difference in mean VOT between children and adults.

c) The 95% confidence interval is $\bar{x}_1 - \bar{x}_2 \pm t^* \sqrt{\frac{s_1^2}{n_1} + \frac{s_2^2}{n_2}}$, where t^* is the upper $(1-C)/2 = 0.025$ critical value for the t distribution with degrees of freedom equal to the smaller of $n_1 - 1 = 9$ and $n_2 - 1 = 19$. Using 9 degrees of freedom we find the value of $t^* = 2.262$, and the confidence interval is

$$-3.67 - (-23.17) \pm 2.262(15.607) = (-15.8, 54.8).$$

Since the interval includes the value 0, we fail to reject H_0 at the 5% level of signficance. We knew this from (a) since the P-value exceeded 5%.

Exercise 7.75

The pooled estimator of σ^2 is

$$s_p^2 = \frac{(n_1-1)s_1^2 + (n_2-1)s_2^2}{n_1+n_2-2} = \frac{(133-1)(5.05)^2 + (162-1)(5.44)^2}{133+162-2} = 27.751.$$

The value of the pooled two-sample t statistic is

$$t = \frac{\bar{x}_1 - \bar{x}_2}{s_p\sqrt{\dfrac{1}{n_1}+\dfrac{1}{n_2}}} = \frac{25.34 - 24.94}{5.268\sqrt{\dfrac{1}{133}+\dfrac{1}{162}}} = 0.649$$

which agrees quite closely with the answer in Example 7.16. When the two standard deviations are close, the numerical values of the two-sample t statistic and the pooled two-sample t statistic are quite similar, as are the conclusions.

The P-value is found by comparing the computed value of t to critical values for the $t(n_1 + n_2 -2) = t(293)$ distribution, and then doubling the upper tail probability because the alternative is two-sided. Table D shows the value of 0.649 does not reach the critical value corresponding to the upper 0.25 tail probability, which is the largest upper tail probability in the table. Doubling this tail area we can see that the P-value exceeds .5. Statistical software can be used to compute the P-value giving the exact result of 0.5168. The data give no evidence of a difference in mean scores between men and women.

SECTION 7.3

OVERVIEW

There are formal inference procedures to compare the standard deviations of two normal populations as well as the two means. The validity of the procedures is seriously affected by nonnormality, and they are not recommended for regular use. The procedures are based on the **F statistic,** which is the ratio of the two sample variances

$$F = \frac{s_1^2}{s_2^2}.$$

If the data consist of independent simple random samples of sizes n_1 and n_2 from two normal populations, then the F statistic has the F distribution, $F(n_1 - 1, n_2 - 1)$, if the two population standard deviations σ_1 and σ_2 are equal.

The power of the pooled two-sample t test can be found using the noncentral t-distribution or a normal approximation to it. The critical value for the significance test, the degrees of freedom and the noncentrality parameter for alternatives of interest are all required for the computation. Power calculations can be useful when comparing alternative designs and assumptions.

GUIDED SOLUTIONS

Exercise 7.83

KEY CONCEPTS - the F test for equality of the standard deviations of two normal populations

The data are from Exercise 7.48. The data consists of measuring the Vitamin C content of two loaves of bread immediately after baking, and the Vitamin C content of two loaves which were stored for three days.

Immediately after baking	47.62	49.79
Three days after baking	21.25	22.34

a) We are interested in testing the hypotheseses $H_0 : \sigma_1 = \sigma_2$ and $H_a : \sigma_1 \neq \sigma_2$. . The two-sided test statistic is the larger variance divided by the smaller variance and if σ_1 and σ_2 are equal has the $F(n_1 - 1, n_2 - 1) = F(2 - 1, 2 - 1) = F(1, 1)$ distribution. From Table E, what value would the ratio of variances have to exceed to reject the null hypothesis at the 5% level of significance? What do you need to do with the significance level when going to the Table E? (Remember it is a two-sided alternative.) What does your answer suggest about the power of the test?

b) To perform the test, first find the two sample variances and take the ratio of the larger to the smaller. Then compare this ratio to the critical value that you obtained in (a). What can be said about the P-value from the table, or using computer software?

Exercise 7.89

KEY CONCEPTS - computing the power of the pooled two-sample t test

Power computations are important when planning a study. You want to ensure you have enough data to detect alternative hypotheses that are of interest to you. Remember, just because a test is not statistically significant doesn't imply that the null hypothesis is true. This situation may occur when the test is based on a small sample size and has low power, or ability, to detect alternatives of interest. Power calculations done before the study help to avoid this problem.

We follow the sequence of steps given in the text, to keep our calculations organized.

1. Specify

a) An alternative that you consider important. Read through the exercise. What alternative (value of $\mu_1 - \mu_2$) is the researcher interested in detecting?

b) What sample sizes are the researchers planning on using?

c) The type I error

d) A guess at the standard deviation σ. This is based on previous data. What value are the researchers going to use?

2. Find the df $= n_1 + n_2 - 2$ and the value t^* leading to rejection.

3. Calculate the noncentrality parameter

$$\delta = \frac{|\mu_1 - \mu_2|}{\sigma\sqrt{(1/n_1)+(1/n_2)}} =$$

4. The power is the probability that a noncentral t random variable with 198 degrees of freedom and noncentrality parameter $\delta = 3.26$ will be less than $t^* = 1.65$. If you do not have software to perform this calculation you can approximate the power as $P(Z > t^* - \delta)$.

$P(Z > t^* - \delta) =$

COMPLETE SOLUTIONS

Exercise 7.83

a) You need to go to the $F(1, 1)$ distribution in Table E and find the critical value corresponding to $0.05/2 = 0.025$ since the test is two-sided. The value in the table is 647.79. The ratio of the larger to smaller variance would need to exceed 647.79 to reject the null hypothesis of equal standard deviations. This suggests that the power of the test will be very low, as this is unlikely to occur unless the population standard deviations are very far apart.

b) For the data collected immediately after baking, the sample standard deviation is 1.53 (mg/100 g), and the sample variance is $(1.53)^2 = 2.3409$. For the data collected three days after baking, the sample standard deviation is 0.771 (mg/100 g), and the sample variance is $(0.771)^2 = 0.5944$. Computing the test statistic

$$F = \frac{\text{larger } s^2}{\text{smaller } s^2} = \frac{2.3409}{0.5944} = 3.94$$

we see that the computed value does not exceed 647.79, so we fail to reject at the 5% level of significance. Using statistical software, the exact upper tail probability (probability of exceeding 3.94 for an $F(1, 1)$ distribution) is 0.2981, and the P-value = $2 \times 0.2981 = 0.5962$.

Exercise 7.89

1.
a) The researcher thinks that the true difference in mean birth weights might be about 300, and a difference this large is clinically important. The value $\mu_1 - \mu_2 = 300$ is an alternative considered important to detect.

b) The sample sizes are $n_1 = 100$ and $n_1 = 100$.

c) The type I error is $\alpha = 0.05$.

d) Our guess for σ is going to be 650.

2. The df $= n_1 + n_2 - 2 = 198$ and the value $t^* = 1.65$.

3. The noncentrality parameter is

$$\delta = \frac{|\mu_1 - \mu_2|}{\sigma\sqrt{(1/n_1) + (1/n_2)}} = \frac{300}{650\sqrt{(1/100) + (1/100)}} = 3.26$$

4. You can approximate the power as

$P(Z > t^* - \delta) = P(Z > 1.65 - 3.26) = P(Z > -1.61) = 1 - .0537 = .9463$.

Using SAS the probability can be computed exactly using the noncentral t distribution. The SAS function which does this is called PROBT and to use it you need to specify t^*, DF, and δ. The SAS "command" is

POWER = 1-PROBT$(t^*$, DF, $\delta) = 1 - $ PROB $(1.65, 198, 3.26) = 0.945925455$.

The normal approximation and the exact calculation agree well in this case. Provided the degrees of freedom is not too small (at least 30 - 40), the agreement will be good and the normal approximation will be sufficient for applied work.

With 100 observations in each group, there is a high probability (almost 95%) that our test will be able to detect a difference in means as large as 300 grams. Said another way, if the difference in means in as large as 300 grams, the probability we will rejects the null hypothesis, which would be the correct decision, is about 95%. The sample sizes are sufficiently large to detect a meaningful difference.

CHAPTER 8

INFERENCE FOR PROPORTIONS

SECTION 8.1

OVERVIEW

In this section, we consider inference about a population proportion p based on the **sample proportion** $\hat{p} = X/n$ obtained from an SRS of size n, where X is the number of "successes" (occurrences of the event of interest) in the sample. If the population is at least 10 times larger than the sample, the individual observations will be approximately independent and X will have a distribution that is approximately binomial $B(n, p)$. Although it is possible to develop procedures for inference about p based on the binomial $B(n, p)$ distribution, these can be awkward to work with because of the discrete nature of the binomial distribution. We therefore restrict ourselves to the case where n is large enough (both $np \geq 10$ and $n(1 - p) \geq 10$) that the normal approximation to the binomial is appropriate. In this case, we can treat $\hat{p}$ as having a distribution that is approximately normal with mean $\mu = p$ and standard deviation $\sigma = \sqrt{p(1-p)/n}$.

An **approximate level C confidence interval** for p is

$$\hat{p} \pm z^* \, \mathrm{SE}_{\hat{p}}$$

where z^* is the upper $(1 - C)/2$ critical value of the standard normal distribution,

$$\text{SE}_{\hat{p}} = \sqrt{\frac{\hat{p}(1-\hat{p})}{n}}$$

is the **standard error** of $\hat{p}$, and $z^*\,\text{SE}_{\hat{p}}$ is the **margin of error.**

Tests of the hypothesis H_0: $p = p_0$ are based on the z–**statistic**

$$z = \frac{\hat{p} - p_0}{\sqrt{\dfrac{p_0(1-p_0)}{n}}}$$

with P–values calculated from the $N(0, 1)$ distribution.

The **sample size** n required to obtain a confidence interval of approximate margin of error m for a proportion is

$$n = \left(\frac{z^*}{m}\right)^2 p^*(1 - p^*)$$

where p^* is a guessed value for the population proportion and z^* is the upper $(1 - C)/2$ critical value of the standard normal distribution. To guarantee that the margin of error of the confidence interval is less than or equal to m no matter what the value of the population proportion may be, use a guessed value of $p^* = 1/2$, which yields

$$n = \left(\frac{z^*}{2m}\right)^2$$

GUIDED SOLUTIONS

Exercise 8.1

KEY CONCEPTS – when to use the normal approximation to the binomial

Recall that the rule of thumb is that the normal approximation to the binomial is appropriate if *both* $np \geq 10$ and $n(1 - p) \geq 10$. These are the conditions that we must check in each of (a) through (d). If we do not know p, we replace p by $\hat{p}$ in these conditions.

Exercise 8.7

KEY CONCEPTS – confidence intervals for a proportion

Recall that an approximate level C confidence interval for p, the true proportion of the month's orders that were shipped on time, is

$$\hat{p} \pm z^* \sqrt{\frac{\hat{p}(1-\hat{p})}{n}}$$

where z^* is the upper $(1 - C)/2$ critical value of the standard normal distribution. From the information given in the problem, provide the values requested below.

n = sample size =

$\hat{p}$ = sample proportion of the month's orders shipped on time =

C = level of confidence requested =

z^* = upper $(1 - C)/2$ critical value of the standard normal distribution =

You will need to use Table D to find z^*. Now substitute these values into the formula for the confidence interval to complete the problem.

$$\hat{p} \pm z^* \sqrt{\frac{\hat{p}(1-\hat{p})}{n}} =$$

Exercise 8.11

KEY CONCEPTS – testing hypotheses about a proportion

Before proceeding, we review some facts. We know that the sampling distribution of any statistic is the distribution of the values of the statistic that we obtain from many, many independent samples. The survey method devised by the market researchers yields a statistic; namely, the proportion of Indiana households in rural areas that we obtain using the method. This statistic will have a sampling distribution. Let p denote the mean of this sampling distribution. This is our population parameter in this problem, and the proportion $\hat{p} = 0.38$ of rural households obtained in the sample actually conducted is an estimate of p.

We are interested in determining if the method used by the market researchers to collect the survey data gives samples that are representative of the state in regard to rural versus urban residence. In the terminology of statistics, this question is equivalent to asking if the survey method devised by the market researchers yields an unbiased estimate of the "true" proportion of Indiana households that are in rural areas, which is 0.36 based on the census data. In other words, is $p = 0.36$? Thus, in terms of a formal test of hypotheses, we want to test

$$H_0: p = 0.36 \text{ vs. } H_a: p \neq 0.36$$

Recall that tests of the hypothesis $H_0: p = p_0$ are based on the z–statistic

$$z = \frac{\hat{p} - p_0}{\sqrt{\dfrac{p_0(1 - p_0)}{n}}}$$

We have already mentioned that $\hat{p} = 0.38$ here. Identify n and p_0 in this example and calculate z.

$$z = \frac{\hat{p} - p_0}{\sqrt{\dfrac{p_0(1 - p_0)}{n}}} =$$

For our hypotheses, the P–value is

$$2P(\text{standard normal random variable} \geq |z|)$$

(Why is the probability multiplied by 2?)

Calculate this P–value and then summarize your conclusions.

$$P\text{–value} = 2P(\text{standard normal random variable} \geq |z|)$$

$$=$$

Conclusions:

Exercise 8.21

KEY CONCEPTS – sample size and margin of error

The sample size n required to obtain a confidence interval of approximate margin of error m for a proportion is

$$n = \left(\frac{z^*}{m}\right)^2 p^*(1 - p^*)$$

where p^* is a guessed value for the population proportion and z^* is the critical value of the standard normal distribution for the desired level of confidence. To apply this formula here, we must determine

m = desired margin of error =

p^* = a guessed value for the population proportion =

C = desired level of confidence =

z^* = the upper $(1 - C)/2$ critical value of the standard normal distribution

   =

From the statement of the exercise, what are these values? Once you have determined them, use the formula to compute the required sample size n.

$$n = \left(\frac{z^*}{m}\right)^2 p^*(1 - p^*) =$$

COMPLETE SOLUTIONS

Exercise 8.1

a) We see that $n\hat{p} = (30)(0.9) = 27 \geq 10$, but $n(1-\hat{p}) = (30)(1 - 0.9) = (30)(0.1) = 3 < 10$. Thus, the normal approximation to the binomial should *not* be used in this case.

b) We see that $n\hat{p} = (25)(0.5) = 12.5 \geq 10$ and $n(1 - \hat{p}) = (25)(1 - 0.5) = (25)(0.5) = 12.5 \geq 10$. Thus, the normal approximation to the binomial can be used in this case.

c) We see that $n\hat{p} = (100)(0.04) = 4 < 10$, so the normal approximation to the binomial should *not* be used in this case.

d) We see that $n\hat{p} = (600)(0.6) = 360 \geq 10$ and $n(1 - \hat{p}) = (600)(1 - 0.6) = (600)(0.4) = 240 \geq 10$. Thus, the normal approximation to the binomial can be used in this case.

Exercise 8.7

The desired information is

n = sample size = 100

$\hat{p}$ = sample proportion of the orders shipped on time = 86/100 = 0.86

C = level of confidence requested = 0.95

z^* = upper $(1 - C)/2$ critical value of the standard normal distribution

= upper 0.025 critical value of the standard normal distribution = 1.96

Substituting these values into the formula for the confidence interval yields

$$\hat{p} \pm z^* \sqrt{\frac{\hat{p}(1 - \hat{p})}{n}} \quad = 0.86 \pm (1.96)\sqrt{\frac{(0.86)(1 - 0.86)}{100}}$$

$$= 0.86 \pm (1.96)(0.0347) = 0.86 \pm 0.068$$

or 0.792 to 0.928.

Exercise 8.11

Here $n = 500$ and $p_0 = 0.36$, so

$$z = \frac{\hat{p} - p_0}{\sqrt{\frac{p_0(1 - p_0)}{n}}} = \frac{0.38 - 0.36}{\sqrt{\frac{0.36(1 - 0.36)}{500}}} = \frac{0.02}{0.0215} = 0.93$$

Thus

$$P\text{-value} = 2P(\text{standard normal random variable} \geq |z|)$$

$$= 2P(\text{standard normal random variable} \geq 0.93)$$

$$= 2(1 - P(\text{standard normal random variable} < 0.93))$$

$$= 2(1 - 0.8238)$$

$$= 0.3524$$

Conclusions: The P-value is not very small, suggesting that the data do not provide convincing evidence against H_0. In other words, the survey method used by the market researchers appears to represent the state in regard to rural versus urban residence.

Exercise 8.21

From the statement of the exercise, we have

$$m = \text{desired margin of error} = 0.05$$

$$p^* = \text{a guessed value for the population proportion} = 0.2$$

$$C = \text{desired level of confidence} = 0.95$$

$$z^* = \text{the upper } (1 - C)/2 \text{ critical value of the standard normal distribution}$$

$$= 1.96$$

The required sample size n is thus

$$n = \left(\frac{z^*}{m}\right)^2 p^*(1 - p^*) = \left(\frac{1.96}{0.05}\right)^2 (0.2)(1 - 0.2) = 245.86$$

which we round up to 246.

SECTION 8.2

OVERVIEW

Confidence intervals and tests designed to compare two population proportions are based on the **difference in the sample proportions** $D = \hat{p}_1 - \hat{p}_2$. The formula for the level C confidence interval is

$$(\hat{p}_1 - \hat{p}_2) \pm z^* \text{SE}_D$$

where z^* is the upper $(1 - C)/2$ standard normal critical value and SE_D is the standard error for the difference in the two proportions computed as

$$\text{SE}_D = \sqrt{\frac{\hat{p}_1(1-\hat{p}_1)}{n_1} + \frac{\hat{p}_2(1-\hat{p}_2)}{n_2}}$$

Significance tests for the equality of the two proportions, H_0: $p_1 = p_2$, use a different standard error for the difference in the sample proportions that is based on a **pooled estimate** of the common (under H_0) value of p_1 and p_2,

$$\hat{p} = \frac{X_1 + X_2}{n_1 + n_2}$$

The test uses the z-statistic

$$z = \frac{\hat{p}_1 - \hat{p}_2}{\text{SE}_{Dp}}$$

where

$$\text{SE}_{Dp} = \sqrt{\hat{p}(1-\hat{p})\left(\frac{1}{n_1} + \frac{1}{n_2}\right)}$$

and P-values are computed using Table A of the standard normal distribution.

GUIDED SOLUTIONS

Exercise 8.27

KEY CONCEPTS - standard errors of proportions, confidence intervals for the difference between two population proportions

The data which give the number of female references that are juvenile (girl) and the number of male references that are juvenile (boy) are reproduced below with the sample sizes for each group.

Gender	n	X(juvenile)
Female	60	48
Male	132	52

a) The proportion of juvenile references for females and its standard error is a question about a single population, females, and does not require any comparison between the groups. The formulas are the same as in Section 8.1. In the space provided give the proportion of juvenile references for females and males, and their standard errors. The subscripts F and M are used to denote whether a sample proportion, sample size or standard error is from the female or male group, respectively.

Females $\hat{p}_F =$ $SE_{\hat{p}_F} =$

Males $\hat{p}_M =$ $SE_{\hat{p}_M} =$

b) To find the confidence interval, you must first compute the standard error for the difference in the two proportions given by the formula

$$SE_D = \sqrt{\frac{\hat{p}_F(1-\hat{p}_F)}{n_F} + \frac{\hat{p}_M(1-\hat{p}_M)}{n_M}} =$$

The level C confidence interval for p_F - p_M is $(\hat{p}_F - \hat{p}_M) \pm z^*SE_D$, where $\hat{p}_F$ and $\hat{p}_M$ were computed in (a), SE_D was computed above, and z^* is the upper $(1 - C)/2$ standard normal critical value. What is the numerical value of C ? Use it to find z^*. Compute the confidence interval in the space below and briefly summarize what the data show.

$(\hat{p}_F - \hat{p}_M) \pm z^*SE_D =$

Exercise 8.35

KEY CONCEPTS - sample proportions, confidence intervals for the difference between two populations proportions

a) A sample proportion is the number of successes divided by the number of observations. Suppose $\hat{p}_1$ is the proportion of farmers in favor of the program in Tippecanoe County, and $\hat{p}_2$ is the proportion in Benton County. Find the values of $\hat{p}_1$ and $\hat{p}_2$ from the information given.

b) Recall that the formula for the standard error of the confidence interval is given by

$$SE_D = \sqrt{\frac{\hat{p}_1(1-\hat{p}_1)}{n_1} + \frac{\hat{p}_2(1-\hat{p}_2)}{n_2}} =$$

You evaluated $\hat{p}_1$ and $\hat{p}_2$ in (a). What are the numerical values of n_1 and n_2? Now you have all the information to evaluate the standard error above.

c) The level C confidence interval for p_1 - p_2 is $(\hat{p}_1 - \hat{p}_2) \pm z^*SE_D$, where $\hat{p}_1$ and $\hat{p}_2$ were computed in (a), SE_D was computed in (b), and z^* is the upper (1 - C)/2 standard normal critical value. What is the numerical value of C ? Use it to find z^*. Now you can compute the confidence interval below.

$(\hat{p}_1 - \hat{p}_2) \pm z^*SE_D =$

Does this interval give evidence that the opinions differed in the two counties? Interpret your interval.

Exercise 8.37

KEY CONCEPTS - pooled estimate of the proportion, testing equality of two population proportions

a) The overall proportion of farmers who favor the corn checkoff program is found by combining the farmers from the two counties. It is the number in favor from both counties divided by the number sampled from both counties. If X_1 and X_2 are the number who favor the favor the program in Tippecanoe and

Benton County respectively, and n_1 and n_2 are the sample sizes, use the formula below to compute the overall proportion.

$$\hat{p} = \frac{X_1 + X_2}{n_1 + n_2} =$$

b) The standard error for testing that the population proportions favoring the program are the same in two counties is expressed in terms of the overall proportion computed in (a). Evaluate the standard error using the formula below.

$$SE_{Dp} = \sqrt{\hat{p}(1-\hat{p})\left(\frac{1}{n_1} + \frac{1}{n_2}\right)} =$$

c) We are interested in seeing if there is a difference between the two counties. What does this say about the alternative? Is it one–sided or two–sided? Write down the null and alternative hypotheses in terms of the population proportions p_1 and p_2.

d) The z–statistic for testing equality of two population proportions is given by the formula

$$z = \frac{\hat{p}_1 - \hat{p}_2}{SE_{Dp}} =$$

Evaluate the z–statistic above using the value of SE_{Dp} computed in (a) and the values of $\hat{p}_1$ and $\hat{p}_2$ computed in Exercise 8.35. Compute the P–value and give your conclusion. When computing the P–value, remember that the alternative is two–sided.

COMPLETE SOLUTIONS

Exercise 8.27

a) Females

$$\hat{p}_F = 48/60 = 0.800, \qquad SE_{\hat{p}_F} = \sqrt{\frac{\hat{p}_F(1-\hat{p}_F)}{n_F}} = \sqrt{\frac{0.80(1-0.80)}{60}} = 0.0516$$

Males

$$\hat{p}_M = 52/132 = 0.394, \quad SE_{\hat{p}_M} = \sqrt{\frac{\hat{p}_M(1-\hat{p}_M)}{n_M}} = \sqrt{\frac{0.394(1-0.394)}{132}} = 0.0425$$

b) The standard error for the difference in the two proportions is

$$SE_D = \sqrt{\frac{\hat{p}_F(1-\hat{p}_F)}{n_F} + \frac{\hat{p}_M(1-\hat{p}_M)}{n_M}} = \sqrt{\frac{0.80(1-0.80)}{60} + \frac{0.394(1-0.394)}{132}}$$

$$= 0.0669$$

For the 95% confidence interval, $C = 0.95$ and we need the upper $(1 - C)/2 = (1 - 0.95)/2 = 0.025$ standard normal critical value. This can be obtained most easily from the bottom row of Table D, with $z* = 1.96$. The confidence interval is given by

$$\hat{p}_F - \hat{p}_M \pm z*SE_D = (0.80 - 0.394) \pm 1.96(0.0669) = 0.406 \pm 0.1311$$

$$= (0.275, 0.537).$$

The difference in proportions is between 0.275 and 0.537. Female references are 40.6% more likely to be juvenile than male references, with a 95% margin of error of 13.1%.

Exercise 8.35

a) In Tippecanoe County, there were 263 in favor of the program among the $n_1 = 263 + 252 = 515$ farmers sampled. (The number sampled is the number in favor plus the number opposed.) So the value of $\hat{p}_1 = 263/315 = 0.511$. Similarly, $n_2 = 260 + 377 = 637$, and $\hat{p}_2 = 260/637 = 0.408$.

b) The value of the standard error for the confidence interval is

$$SE_D = \sqrt{\frac{\hat{p}_1(1-\hat{p}_1)}{n_1} + \frac{\hat{p}_2(1-\hat{p}_2)}{n_2}} = \sqrt{\frac{0.511(1-0.511)}{515} + \frac{0.408(1-0.408)}{637}}$$

$$= 0.0294$$

c) For the 99% confidence interval, $C = 0.99$ and we need the upper $(1 - C)/2 = (1 - 0.99)/2 = 0.005$ standard normal critical value. This can be obtained most easily from the bottom row of Table D, with $z* = 2.576$. The confidence interval is given by

$$(\hat{p}_1 - \hat{p}_2) \pm z*SE_D = (0.511 - 0.408) \pm 2.576(0.0294) = 0.103 \pm 0.076$$

$$= (0.027, 0.179)$$

Since the interval doesn't include 0, the evidence is that the opinions differed, with more farmers in Tippecanoe in favor of the program. We estimate 10.3% more farmers from Tippecanoe favor the program, with a 99% margin of error of 7.6%

Exercise 8.37

a) Using the data from Exercise 8.35, we have $X_1 = 263$, $X_2 = 260$, $n_1 = 263 + 252 = 515$, and $n_2 = 260 + 377 = 637$. The overall proportion is

$$\hat{p} = \frac{X_1 + X_2}{n_1 + n_2} = \frac{263 + 260}{515 + 637} = 0.454$$

b) The standard error used to test equality of the population proportions is

$$SE_{Dp} = \sqrt{\hat{p}(1-\hat{p})\left(\frac{1}{n_1} + \frac{1}{n_2}\right)} = \sqrt{0.454(1-0.454)\left(\frac{1}{515} + \frac{1}{637}\right)} = 0.0295$$

c) The hypotheses are H_0: $p_1 = p_2$ and H_a: $p_1 \neq p_2$. The alternative is two–sided, since we are interested in differences in either direction.

d) The numerical value of the z-statistic for testing equality of two population proportions is

$$z = \frac{\hat{p}_1 - \hat{p}_2}{SE_{Dp}} = \frac{0.511 - 0.408}{0.0295} = 3.49$$

We need to first use table A to find the probability of getting a z-statistic at least this large, and then this probability must be doubled to obtain the P-value, since the alternative is two–sided.

$P(Z > 3.49) = 0.0002$ from Table A. So the P-value is 0.0004. There is very strong evidence that the proportion in favor of the program differs between the two counties. The proportion in favor of the program is higher in Tippecanoe.

CHAPTER 9

INFERENCE FOR TWO–WAY TABLES

OVERVIEW

There are two common models that generate data that can be summarized in a two–way table. In the first model, independent SRSs are drawn from c populations and each observation is classified according to a categorical variable that has r possible values. In the table, the c populations form the columns, and the values of categorical classification variable form the rows. The null hypothesis is that the distributions of the row categorical variable values are the same for each of c populations. In the second model, an SRS is drawn from a single population, and the observations are cross–classified according to two categorical variables having r and c possible values. In this model, the null hypothesis is that the row and column variables are independent. When one of the variables is an explanatory variable and the other is a response, the explanatory variable is used to form the columns of the table and the response forms the rows.

A test of the null hypothesis is carried out using X^2 in both models. Although the data are generated differently, the question in both cases is quite similar: Are the distributions of the column variables the same? The cell counts are compared to the **expected cell counts** under the null hypothesis. The expected cell counts are computed using the formula

$$\text{expected count} = \frac{\text{row total x column total}}{n}$$

where n is the total number of observations.

The **chi–square statistic** is used to test the null hypothesis by comparing the observed counts with the expected counts

$$X^2 = \sum \frac{(\text{observed - expected})^2}{\text{expected}}$$

When the null hypothesis is true, the distribution of X^2 is approximately χ^2 with $(r - 1)(c - 1)$ degrees of freedom. The P-value is the probability of getting differences between observed and expected counts as large as we did, and is computed as $P(\chi^2 > X^2)$. The use of the χ^2 distribution is an approximation that works well when the average expected count exceeds 5 and all of the individual expected counts are greater than 1. In the special case of the 2 x 2 table, all expected counts should exceed 5 before applying the approximation.

In addition to computing the X^2 statistic, tables or bar charts should be examined to describe the relationship between the two variables.

GUIDED SOLUTIONS

Exercise 9.9

KEY CONCEPTS – testing independence in a 2×2 table

a) For these data, which is the explanatory variable and which is the response? To be an experiment, what needs to be true about the assignment of the explanatory variable to the subjects (review Sections 3.1 and 3.2)? Was this carried out here?

b) Pet ownership defines the two populations. The percentages that you compute should allow comparisons between these two populations. In this case, how does the percentage of patients alive compare for the two categories of pet ownership? Compute the appropriate percentages in the table below and state your preliminary conclusions.

	Pet ownership	
Patient status	No	Yes
Alive		
Dead		

c) In this exercise, was a single SRS drawn from a population, or were independent SRSs drawn from each population? This is important for stating the null hypothesis. Don't just state the hypotheses generally in terms of row and column variables. Be sure to state the hypotheses using the names of the variables in this particular problem.

d) The easiest way to find the value of the chi–square statistic is to use statistical software. If you don't have access to statistical software, the computations must be done by hand. The first step is to determine the expected counts for each of the cells. Remember that the formula is

$$\text{expected count} = \frac{\text{row total x column total}}{n}$$

It would be easiest to add the row totals, the column totals, and the value of n to the original table below before computing the expected counts. In the table below, fill in the numerical values of the expected counts below the counts in each of the cells.

| | Pet ownership | | |
Patient status	No	Yes	Total
Alive	28	50	
Dead	11	3	
Total			

The chi-square statistic compares the observed counts to the expected counts using the formula below. What is the value of X^2 in this problem? Remember, there are four terms in the sum, one corresponding to each of the cells.

$$X^2 = \sum \frac{(\text{observed - expected})^2}{\text{expected}} =$$

X^2 has an approximate χ^2 distribution with degrees of freedom $(r - 1)(c - 1)$, provided all the expected counts are 5 or greater. What are r and c in this case? Verify that the approximation can be applied.

Now use Table F to determine the *P*-value. It is the area to the right of the value of X^2 using the line in the table corresponding to the degrees of freedom you computed.

e) Are these observational data, or do they form an experiment? This determines the type of conclusion you can draw. Is there a cause and effect relationship or just an association? Can the effects be attributed to confounding variables? Review these ideas from Sections 3.1 and 3.2 and apply them to this data set.

Exercise 9.17

KEY CONCEPTS – testing independence of row and column variables, describing the association

The details of hand calculations were presented in Exercise 9.9. In this problem, we present the output using computer software and concentrate on the analysis and interpretation. The output gives the details of the calculation of X^2 using MINITAB. The table of counts is reproduced as part of the computer output, with the expected counts printed directly below for each cell.

```
Expected counts are printed below observed counts

   Alcohol          Nicotine (milligrams/day)
(ounces/day)     None      1-15     16 or more   Total

None              105         7         11         123
                82.73     17.69      22.59

0.01-0.10          58         5         13          76
                51.12     10.93      13.96

0.11-0.99          84        37         42         163
               109.63     23.44      29.93

1.00 or more       57        16         17          90
                60.53     12.94      16.53

Total             304        65         83         452

ChiSq =  5.997 +   6.458 +   5.944 +   0.927 +   3.217 +
         0.065 +   5.991 +   7.844 +   4.866 +   0.206 +
0.722 +  0.014 = 42.252

df = _____, p = _____
```

Below the table of observed and expected counts, the computation of

$$X^2 = \sum \frac{(\text{observed} - \text{expected})^2}{\text{expected}}$$

is provided, by giving the value of $\frac{(\text{observed} - \text{expected})^2}{\text{expected}}$ for each of the 12
cells, going across the table in rows. For example, the second component of X^2
is 6.458 and is computed as $\frac{(7 - 17.69)^2}{17.69} = 6.458$ (this corresponds to the second
column in the first row). The cells for which the component of X^2 is large are
cells for which the agreement between the observed count and the expected
count under the null hypothesis of independence is not close. Before
proceeding further with the analysis, make sure you understand how the totals,
the expected counts and the X^2 were computed. Review Exercise 9.9 if you are
having difficulty reproducing any of the numbers in the output.

The remainder of the computer output gives the degrees of freedom associated
with X^2 and the P-value. The answers have been omitted. Make sure you can
determine the degrees of freedom and know how to use Table F to find the
P-value.

At this point the formal statistical analysis tells us that there is an association
between alcohol intake prior to pregnancy and nicotine intake during
pregnancy. The extremely large X^2 and the small P-value associated with it
indicate this association cannot be explained by chance. Describing the nature
and strength of the association is the next part of the analysis. This can be done
by computing conditional distributions (looking at column or row percents) and
presenting the results in a table or by providing a bar graph to compare these
conditional distributions.

When one variable can be considered the explanatory variable and the other the
response, then you would usually examine the conditional distribution of the
response for each level of the explanatory variable. When there is no
explanatory-response distinction, then you can look at either or both sets of
conditional distributions. Is there an explanatory-response variable distinction
in this problem?

In the table on the next page, give the conditional distribution of alcohol intake
for each level of nicotine intake.

Alcohol	Nicotine (milligrams/day)		
(ounces/day)	None	1-15	16 or more
None			
0.01-0.10			
0.11-0.99			
1.00 or more			
	100%	100%	100%

COMPLETE SOLUTIONS

Exercise 9.9

a) For this data, the explanatory variable is pet ownership and the response is patient status. The experimenter did not intervene and assign the subjects to own or not own a pet. They just observed pet ownership status in the data. That makes these data observational data, and there is a possibility that differences in patient status may be a result of confounding variables, not necessarily to pet ownership.

b) Since pet ownership is the explanatory variable, we want to look at the distribution of patient status for each category of pet ownership. In this case the column percentages are of interest.

	Pet ownership	
Patient status	No	Yes
Alive	72%	94%
Dead	28%	6%
	100%	100%

There are large differences in the percentages. The death rate among those not owning pets appears to be higher.

c) The data is viewed as a sample from the population of patients with CHD who are classified according to two categorical variables. In this model, the null hypothesis is that pet ownership and patient status are independent. The alternative is that they are not independent. The alternative is general and does not imply any particular type of dependence or relationship between the variables.

d) The original table is reproduced below with the expected counts for each cell listed below each of the counts. The column totals, row totals and n have been added to the table as well. The expected counts were computed as follows:

(Alive, No)   expected count $= \dfrac{\text{row total x column total}}{n} = \dfrac{78 \times 39}{92} = 33.07$

(Alive, Yes)   expected count $= \dfrac{78 \times 53}{92} = 44.93$

(Dead, No)   expected count $= \dfrac{14 \times 39}{92} = 5.93$

(Dead, Yes)   expected count $= \dfrac{14 \times 53}{92} = 8.07$

	Pet ownership		
Patient status	No	Yes	
Alive	28	50	78
	33.07	44.93	
Dead	11	3	14
	5.93	8.07	
	39	53	92

The numerical value of X^2 is computed using the formula

$$X^2 = \sum \frac{(\text{observed} - \text{expected})^2}{\text{expected}}$$

$$= \frac{(28 - 33.07)^2}{33.07} + \frac{(50 - 44.93)^2}{44.93} + \frac{(11 - 5.93)^2}{5.93} + \frac{(3 - 8.07)^2}{8.07}$$

$$= 0.777 + 0.572 + 4.33 + 3.19 = 8.87.$$

Since r and c are both 2, the degrees of freedom are $(2-1)(2-1) = 1$. All the expected counts exceed 5 so the χ^2 approximation can be used. From the first row of Table F (corresponding to 1 df) we see that the P-value is between 0.0025 and 0.005. (Statistical software gives $P(\chi^2 > 8.87) = 0.003$ as the P-value).

e) Since this is observational data, you need to make sure that when you state your conclusions you do not imply that there is a cause and effect relationship.

In this case, pet ownership and patient status are associated. One–year survival rates are higher for those who own a pet. But remember that there could be confounding variables. For example, perhaps those who own pets tend to have different lifestyles. It's possible that they have more activities and interests than those who do not own pets and that this is what is helping to improve their chances of survival. Certainly, walking your dog may be helpful in improving your survival rate if you have CHD, and dog owners may be more likely to take walks (of course, taking a walk without a dog may be just as beneficial).

This study didn't show that buying a pet upon your discharge from the coronary care unit would improve your survival. To show this, those leaving the coronary unit would have to be divided into two groups at random. Those in the first group would be given pets and those in the second group would not be given pets. The survival rates of the two groups after one year would then be compared. Of course, although this experiment would show cause and effect, there are obvious practical difficulties in carrying it out.

Exercise 9.17

The computation of X^2 is described in the guided solution. $X^2 = 42.252$ and has $(r-1)(c-1) = (4-1)(3-1) = 6$ degrees of freedom. Going to Table F and using the line corresponding to 6 degrees of freedom, we find the largest entry is 24.10, corresponding to a upper tail area of 0.0005. Since 42.252 is larger than this value, we conclude that the P-value $= P(\chi^2 > 42.252) < 0.0005$.

The table below gives the conditional distribution of the percent alcohol intake for each of the three levels of nicotine intake. The percentages were rounded to whole numbers to make the table easier to read but also made the first two columns add to 101%.

Alcohol	Nicotine (milligrams/day)		
(ounces/day)	None	1-15	16 or more
None	35%	11%	13%
0.01-0.10	19%	8%	16%
0.11-0.99	28%	57%	51%
1.00 or more	19%	25%	20%

The rows that have the biggest discrepancies in percentages are those that give the greatest contribution to X^2. (Verify this for yourself by examining the values of $\dfrac{(\text{observed - expected})^2}{\text{expected}}$ for the 12 cells provided in the guided solutions.)

Most of the association between nicotine and alcohol intake can be explained by the differences in distribution of alcohol intake found in the first and third rows. Those that don't smoke have a much greater percentage who don't drink than either the light or heavy smoking categories. This percentage is approximately three times as high as either of the two smoking categories. If we view the alcohol intake of 0.11-0.99 as "moderate" drinking, then those that smoke (in either smoking category) are much more likely to have had a moderate daily intake of alcohol before pregnancy than those who didn't smoke. The percentage who drink moderate amounts of alcohol is almost twice as high for those who smoke than for nonsmokers. The heaviest and lightest drinking category seem to contain similar percentages for all categories of smokers. So a simple summary of the association would be that smoking during pregnancy seems to be associated with a higher frequency of moderate alcohol intake prior to pregnancy and that not smoking during pregnancy is associated with a higher frequency of no alcohol intake prior to pregnancy. The table giving the conditional distributions can also be summarized in a bar graph as was done in Chapter 2 when summarizing relations in categorical data. The bar graph contains the same information as in the table, but the visual format often makes the relationship easier to see.

Comparision of Alcohol Intake Prior to Pregnancy for Three Categories of
Nicotine Intake During Pregnancy

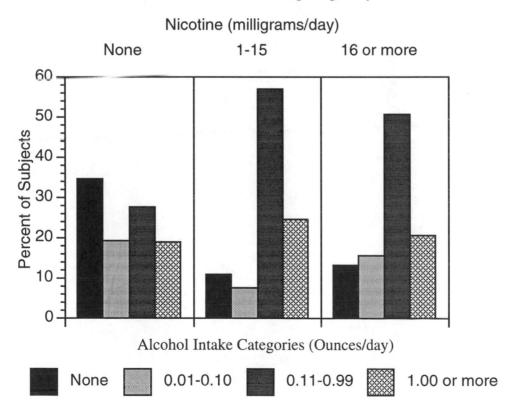

CHAPTER 10

INFERENCE FOR REGRESSION

OVERVIEW

The statistical model for **simple linear regression** is

$$y_i = \beta_0 + \beta_1 x_i + \varepsilon_i$$

where $i = 1, 2,\ldots, n$. The deviations ε_i are assumed to be independent and normally distributed with mean 0 and standard deviation σ. The **parameters** of the model are the intercept β_0, the slope β_1, and σ. β_0 and β_1 are estimated by the slope b_0 and intercept b_1 of the **least-squares regression line**. Given n observations on an explanatory variable x and a response variable y,

$$(x_1, y_1), (x_2, y_2),\ldots, (x_n, y_n),$$

recall that the formula for the slope and intercept of the least-squares regression line are

$$b_1 = r\frac{s_y}{s_x}$$

and

$$b_0 = \bar{y} - b_1 \bar{x},$$

where r is the correlation between y and x, $\bar{y}$ is the mean of the y observations, s_y is the standard deviation of the y observations, $\bar{x}$ is the mean of the x observations, and s_x is the standard deviation of the x observations. The standard deviation σ is estimated by

$$s = \sqrt{\frac{\sum e_i^2}{n-2}}$$

where the e_i are the **residuals**

$$e_i = y_i - \hat{y}_i$$

and

$$\hat{y}_i = b_0 + b_1 x_i.$$

b_0, b_1, and s are usually calculated using a calculator or statistical software.

A **level C confidence interval** for β_1 is

$$b_1 \pm t^* SE_{b_1}$$

where t^* is the upper $(1-C)/2$ critical value for the $t(n-2)$ distribution and

$$SE_{b_1} = \frac{s}{\sqrt{\sum (x_i - \bar{x})^2}}$$

is the standard error of the slope b_1. SE_{b_1} is usually computed using a calculator or statistical software. (The formula above is actually given in Section 10.2 of the text, but reproduced here for continuity of exposition. If your class is not covering Section 10.2, you need not worry about the formula.) The **test of the hypothesis** $H_0: \beta_1 = 0$ is based on the t statistic

$$t = \frac{b_1}{SE_{b_1}}$$

with P-values computed from the $t(n-2)$ distribution. There are similar formulas for confidence intervals and tests for β_0, but using the standard error of the intercept b_0

$$SE_{b_0} = s\sqrt{\frac{1}{n} + \frac{\bar{x}^2}{\sum(x_i - \bar{x})^2}}$$

in place of SE_{b_1}. As with SE_{b_1}, SE_{b_0} is usually computed using a calculator or statistical software. (The formula above is actually given in Section 10.2 of the text, but it is reproduced here for continuity of exposition. If your class is not covering Section 10.2, you need not worry about the formula.)   Note that inferences for the intercept are meaningful only in special cases.

The **estimated mean response** for the subpopulation corresponding to the value x^* of the explanatory variable is

$$\hat{\mu}_y = b_0 + b_1 x^*.$$

A **level C confidence interval for the mean response** is

$$\hat{\mu}_y \pm t^* SE_{\hat{\mu}}$$

where t^* is the upper $(1-C)/2$ critical value for the $t(n-2)$ distribution and

$$SE_{\hat{\mu}} = s\sqrt{\frac{1}{n} + \frac{(x^* - \bar{x})^2}{\sum(x_i - \bar{x})^2}}.$$

$SE_{\hat{\mu}}$ is usually computed using a calculator or statistical software. (The formula above is actually given in Section 10.2 of the text, but it is reproduced here for continuity of exposition. If your class is not covering Section 10.2, you need not worry about the formula.)

The **estimated value of the response variable** y for a future observation from the subpopulation corresponding to the value x^* of the explanatory variable is

$$\hat{y} = b_0 + b_1 x^*.$$

A **level C prediction interval** for the estimated response is

$$\hat{y} \pm t^* SE_{\hat{y}}$$

where t^* is the upper $(1 - C)/2$ critical value for the $t(n-2)$ distribution and

$$SE_{\hat{y}} = s \sqrt{1 + \frac{1}{n} + \frac{(x^* - \bar{x})^2}{\sum (x_i - \bar{x})^2}}.$$

$SE_{\hat{y}}$ is usually computed using a calculator or statistical software. (The formula above is actually given in Section 10.2 of the text, but it is reproduced here for continuity of exposition. If your class is not covering Section 10.2, you need not worry about the formula.)

The **ANOVA table** for a linear regression gives the total sum of squares SSM for the model; the total sum of squares SSE for error; the total sum of squares SST for all sources of variation, the degrees of freedom DFM, DFE, and DFT for these sums of squares; the mean squares MSM for the model; and the mean squares MSE for error. The total sums of squares are

$$SSM = \sum (\hat{y}_i - \bar{y})^2$$

$$SSE = \sum (y_i - \hat{y}_i)^2$$

$$SST = \sum (y_i - \bar{y})^2$$

and the degrees of freedom are

$$DFM = 1$$

$$DFE = n - 2$$

$$DFT = n - 1.$$

The mean sums of squares MS are defined by the relation

$$MS = \frac{\text{sum of squares}}{\text{degrees of freedom}}.$$

The ANOVA table usually has a form like the following.

Source	Degrees of freedom	Sum of squares	Mean square	F
Model	DFM	SSM	MSM = SSM/DFM	MSM/MSE
Error	DFE	SSE	MSE = SSE/DFE	
Total	DFT	SST		

The **ANOVA F statistic** is the ratio MSM/MSE and is used to test H_0: $\beta_1 = 0$, versus the two-sided alternative. Under H_0, this statistic has an $F(1, n - 2)$ distribution which can be used to compute P-values using Table E.

When the variables y and x are jointly normal, the sample correlation is an estimate of the population correlation ρ. The test of H_0: $\rho = 0$ is based on the t **statistic**

$$t = \frac{r\sqrt{n-2}}{\sqrt{1-r^2}}$$

which has a $t(n - 2)$ distribution under H_0. This test statistic is numerically identical to the t statistic used to test H_0: $\beta_1 = 0$.

The **square of the sample correlation** can be written as

$$r^2 = \text{SSM/SST}$$

and is interpreted as the proportion of the variability in the response variable y that is explained by the explanatory variable x in the simple linear regression.

SAMPLE PROBLEMS

GUIDED SOLUTIONS

Exercise 10.19

KEY CONCEPTS - scatterplots, linear regression, significance tests for the slope, prediction intervals

a) What is the explanatory variable and what is the response variable here? Remember that in the scatterplot, the horizontal axis represents the explanatory variable and the vertical axis the response variable. Sketch your scatterplot in the space provided below. Are there any outliers or unusual observations in the plot?

b) If possible, use statistical software to compute the equation of the least-squares line. If you do not have access to statistical software, let x be heart rate HR and y oxygen uptake V02, and compute the following sample quantities from the data.

$$s_x =$$
$$\bar{x} =$$
$$s_y =$$
$$\bar{y} =$$
$$r = \text{the correlation between } x \text{ and } y =$$

From these sample quantities, compute

$$b_1 = r\frac{s_y}{s_x} =$$

and

$$b_0 = \bar{y} - b_1\bar{x} =$$

c) We wish to test the hypothesis $H_0: \beta_1 = 0$. Should H_a be one- or two-sided? Is there anything in the problem suggesting that we are interested in showing the slope is positive? Negative? Or are we just interested in whether or not the slope is 0?

The actual test is best done using statistical software. If you do not have access to statistical software, you will need to compute the residuals $e_i = y_i - \hat{y}_i$ for the data, then compute

$$s = \sqrt{\frac{\sum e_i^2}{n-2}} =$$

and

$$SE_{b_1} = \frac{s}{\sqrt{(x_i - \bar{x})^2}} = \frac{s}{\sqrt{(n-1)s_x^2}} =$$

and finally

$$t = \frac{b_1}{SE_{b_1}} =$$

Use Table D (critical values of the t distribution) to determine the P-value. What are the appropriate degrees of freedom? Explain in plain English what you would conclude from this test.

d) These 95% prediction intervals are most easily computed using statistical software. Minitab, for example, will compute prediction intervals for you. Write these intervals below.

To compute these prediction intervals by hand, we recall the relevant formulas. To compute a 95% prediction interval for the estimated response at the value x^* of the explanatory variable you will need to compute the following

$$\hat{y} \pm t^* SE_{\hat{y}}$$

where

$$\hat{y} = b_0 + b_1 x^*.$$

and t^* is the upper 0.025 critical value for the $t(n - 2) = t(17)$ distribution (which from Table D is 2.11) and

$$SE_{\hat{y}} = s\sqrt{1 + \frac{1}{n} + \frac{(x^* - \bar{x})^2}{\sum(x_i - \bar{x})^2}}.$$

b_0, b_1, and s have already been computed in (b) and (c). Also notice that in the formula for $SE_{\hat{y}}$, $\sum(x_i - \bar{x})^2 = (n - 1)s_x^2$. To further clarify how the computations go, we give the calculations for the case $x^* = 95$.

Recall from (b) that $b_1 = 0.03866$ and $b_0 = -2.805$, and from (c) $s = 0.1205$. We find for $x^* = 95$

$$\hat{y} = b_0 + b_1 x^* = -2.805 + 0.03866(95) = 0.8677$$

$$SE_{\hat{y}} = s\sqrt{1 + \frac{1}{n} + \frac{(x^* - \bar{x})^2}{\sum(x_i - \bar{x})^2}} \quad = s\sqrt{1 + \frac{1}{n} + \frac{(x^* - \bar{x})^2}{(n-1)s_x^2}}$$

$$= (0.1205)\sqrt{1 + \frac{1}{19} + \frac{(95 - 107)^2}{(18)(11.8275)^2}}$$

$$= (0.1205)1.0535$$

$$= 0.1269$$

Now the upper 0.025 critical value for the $t(n-2) = t(17)$ distribution from Table D is $t^* = 2.11$ so our 95% prediction interval is

$$\hat{y} \pm t^* SE_{\hat{y}} = 0.8677 \pm 2.11(0.1269) = 0.8677 \pm 0.2678.$$

Now try the calculations for $x^* = 110$ on your own.

e) In determining whether the researchers should use the predicted V02 in place of the measured V02 for this individual consider the following. From the scatterplot in (a), how close are the points to lying perfectly on a straight line? What does the P-value of the test in (c) indicate about using the least-squares regression line for prediction? What is the margin of error for the 95% prediction intervals found in (d)? Do the observed values of V02 for values of HR near 95 and 110 fall inside or outside the prediction intervals? Should the answer depend on the margin of error for prediction that the researchers can tolerate?

Exercise 10.27

KEY CONCEPTS - the ANOVA table, the ANOVA F=statistic, the square of the correlation

a) The ANOVA table is most easily computed using statistical software. Copy your ANOVA table in the space below.

Source	Degrees of freedom	Sum of squares	Mean square	F
Model				
Error				
Total				

To compute the sums of squares by hand, recall the relevant formulas.

Sum of squares for the model = SSM = $\sum (\hat{y}_i - \bar{y})^2$

Sum of squares for error = SSE = $\sum (y_i - \hat{y}_i)^2$

Sum of squares total = SST = $\sum (y_i - \bar{y})^2 = (n-1)s_y^2$

To simplify things, we note that

$$SSE = \sum (y_i - \hat{y}_i)^2 = (n-2)s^2 =$$

$$SST = \sum (y_i - \bar{y})^2 = (n-1)s_y^2 =$$

where s_y and s were computed in (b) of Exercise 10.19. One then can compute

SSM = SST – SSE =

You will need to compute the degrees of freedom.

Degrees of freedom for SSM = DFM = 1

Degrees of freedom for SSE = DFE = $n - 2 =$

Degrees of freedom for SST = DFT = $n - 1 =$

Finally, the mean sum of squares MS is calculated from the relation

$$MS = \frac{\text{sum of squares}}{\text{degrees of freedom}}$$

and $F = MSM/MSE =$

After computing these quantities, fill in the ANOVA table above.

b) What parameter in the linear regression model is tested by the ANOVA $F=$ statistic? If you are not sure, you may wish to refer to the chapter overview to refresh your memory about what null hypothesis is tested by the ANOVA $F=$statistic.

Interpret this hypothesis in terms of the relation between HR and V02.

c) Recall, that the ANOVA F statistic is the ratio MSM/MSE, and you computed its value in (a). It is used to test $H_0: \beta_1 = 0$ versus the two-sided alternative. Under H_0, this statistic has an $F(1, n - 2)$ distribution. What is n here? P-values can be computed using Table E (refer to Chapter 7 for more on the F-distribution).

d) Recall that the $t=$ statistic calculated in Exercise 10.19 was $t = 16.1$. Refer to (a) for the value of F.

e) Recall that the square of the sample correlation can be written as

$$r^2 = SSM/SST$$

and is interpreted as the proportion of the variation in the response variable y (oxygen uptake, here) that is explained by the explanatory variable x (heart rate, here) in the simple linear regression. Now compute r^2.

Exercise 10.29

KEY CONCEPTS - inferences for the correlation r

a) Recall that when the variables y and x are jointly normal, the sample correlation r is an estimate of the population correlation ρ. The test of H_0: $\rho = 0$ is based on the $t=$ statistic

$$t = \frac{r\sqrt{n-2}}{\sqrt{1-r^2}} =$$

Complete the calculation of t.

b) Recall that the t statistic in (a) has a $t(n-2)$ distribution under H_0. You can use Table D to estimate the P-value. Remember, this is a test of a two-sided hypothesis.

What do you conclude?

COMPLETE SOLUTIONS

Exercise 10.19

a) Since we are interested in whether HR can be used to predict V02, we treat HR as the explanatory variable and V02 as the response. A scatterplot of HR and V02 is given below.

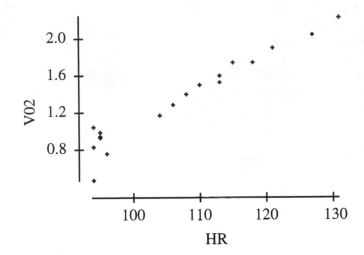

There are no outliers or other unusual points in the plot.

b) Using statistical software, we find the equation of the least-squares regression line is

$$HR = -2.8044 + 0.038652V02$$

For those doing the calculations by hand, we find the following values of the sample statistics:

$$s_x = 11.8275$$
$$\bar{x} = 107$$
$$s_y = 0.4719$$
$$\bar{y} = 1.33142$$
$$r = \text{the correlation between } x \text{ and } y = 0.969$$

From these sample quantities, compute

$$b_1 = r\frac{s_y}{s_x} = (0.969)\frac{0.4719}{11.8275} = 0.03866$$

and

$$b_0 = \bar{y} - b_1\bar{x} = 1.33142 - (0.03866)(107) = -2.805$$

which, except for round-off error, agrees with the values found with statistical software.

c) In this problem, we are simply interested in determining whether or not the slope is 0. This suggests that we should test the hypotheses

$$H_0: \beta_1 = 0$$

$$H_a: \beta_1 \neq 0$$

This test would be based on the test statistic $t = \dfrac{b_1}{SE_{b_1}}$. Using statistical software, we find that $t = 16.1$ and that the P-value of our (two-sided) test is less than 0.001, which is obtained from the critical values of the $t(n-2) = t(17)$ distribution in Table D. Because the test is two-sided, we must double the tail probability given in the Table to get the P-value.

If you do the calculations by hand, compute the residuals $e_i = y_i - \hat{y}_i$ for the data. Then one finds

$$s = \sqrt{\frac{e_i^2}{n-2}} = \sqrt{\frac{0.2468}{17}} = 0.1205$$

and

$$SE_{b_1} = \frac{s}{\sqrt{(x_i - \bar{x})^2}} = \frac{s}{\sqrt{(n-1)s_x^2}} = \frac{0.1205}{\sqrt{(18)(11.8275)^2}} = 0.0024$$

and finally

$$t = \frac{b_1}{SE_{b_1}} = \frac{0.03866}{0.0024} = 16.1$$

Use Table D (critical values of the $t=$ distribution) to determine the P-value. Here we are interested in the $t(n-2) = t(17)$ distribution, and we find, the P-value is less than 0.001 (we must double the tail probability in Table D since we have a two-sided hypothesis).

We conclude that there is strong evidence against the null hypothesis that there is not a straight = line relation between HR and V02. However, even though the P-value is very small, this test does not tell us if the equation of the least-squares regression line will yield accurate predictions of V02 from the values of HR. All we can really conclude is that predictions using the equation of the least-squares regression line will be much better than simply predicting V02 to be the mean of the observed values, regardless of the value of HR. The issue of the accuracy of the prediction using the least-squares regression line is further explored in (d).

d) Using statistical software, we find the following.

For HR = 95, we predict V02 to be 0.8676, and a 95% prediction interval for V02 is (0.5998, 1.1354) or 0.8676 ± 0.2678.

For HR = 110, we predict V02 to be 1.4474, and a 95% prediction interval for V02 is (1.1861, 1.7086) or 1.4474 ± 0.2613.

For those doing the calculations by hand, the 95% prediction interval for $x^* = 95$ was given in the guided solution. For $x^* = 110$, we find

$$\hat{y} = b_0 + b_1 x^* = -2.805 + (0.03866)(110) = 1.4476$$

$$SE_{\hat{y}} = s\sqrt{1 + \frac{1}{n} + \frac{(x^* - \bar{x})^2}{\hat{A}(x_i - \bar{x})^2}} = s\sqrt{1 + \frac{1}{n} + \frac{(x^* - \bar{x})^2}{(n-1)s_x^2}}$$

$$= (0.1205)\sqrt{1 + \frac{1}{19} + \frac{(110 - 107)^2}{(18)(11.8275)^2}}$$

$$= (0.1205)1.0277$$

$$= 0.1238$$

Now the upper 0.025 critical value for the $t(n - 2) = t(17)$ distribution from Table D is $t^* = 2.11$ so our 95% prediction interval is

$$\hat{y} \pm t^* SE_{\hat{y}} = 1.4476 \pm 2.11(0.1238) = 1.4476 \pm 0.2612$$

e) Although from (c) we saw that the P-value of the test of the hypotheses

$$H_0: \beta_1 = 0$$

$$H_a: \beta_1 \neq 0$$

is very small, this test does not tell us if the equation of the least-squares regression line will yield accurate predictions of V02 from the values of HR. The scatterplot in (a) shows that the straight = line relation is good but far from perfect; i.e., there is some variation of the points around a straight line. This is further supported by the results of (d). For HR = 95, we predict V02 to be 0.8676 (which is below all the observed values at HR = 95), and a 95% prediction interval for V02 is 0.8676 ± 0.2678 or (0.5998, 1.1354). Nearly all the observed values for HR = 94, 95, and 96 fall in this interval. For HR = 110, we predict V02 to be 1.4474, and a 95% prediction interval for V02 is 1.4474 ± 0.2613 or (1.1861, 1.7086). All the observed values for HR =106 to HR = 113 fall in this interval. It would appear that predictions using the least-squares regression line are not useful for distinguishing between changes in HR of 1 or 2 but are useful for distinguishing between values of HR that differ by about 8. Ultimately, whether the researchers should use the predicted V02 in place of the measured V02 depends on how accurately the value of V02 needs to be known. The 95% prediction intervals have margins of error of about 0.26. Is such a margin of error too large, or is it adequate for the purposes of the researchers? This is a question the researchers, not the statistician, need to answer.

Exercise 10.27

a) From statistical software, we get the following ANOVA table.

Source	Degrees of freedom	Sum of squares	Mean square	F
Model	1	3.7619	3.7619	259.27
Error	17	0.2467	0.0145	
Total	18	4.0085		

If the calculations are done by hand, we find

$$\text{SSE} = \sum (y_i - \hat{y}_i)^2 = (n-2)s^2 = (17)(0.1205)^2 = 0.2468$$

$$\text{SST} = \sum (y_i - \bar{y})^2 = (n-1)s_y^2 = (18)(0.4719)^2 = 4.0084$$

$$\text{SSM} = \text{SST} - \text{SSE} - 4.0084 - 0.2468 = 3.7616$$

Degrees of freedom for SSM = DFM = 1

Degrees of freedom for SSE = DFE = $n - 2 = 17$

Degrees of freedom for SST = DFT = $n - 1 = 18$

MSM = SSM/DFM = 3.7616/1 = 3.7616

MSE = SSE/DFE = 0.2468/17 = 0.0145

and F = MSM/MSE = 3.7616/0.0145 = 259.42.

Except for round-off error, this agrees with the values found using computer software.

b) The ANOVA F statistic is the ratio MSM/MSE and is used to test the null hypothesis H_0: $\beta_1 = 0$, where β_1 is the slope of the regression line in our linear regression model. In practical terms, this tests whether there is any evidence of a straight = line relation between HR and V02.

c) Under H_0, the $F=$ statistic has an $F(1, n - 2) = F(1, 17)$ distribution. In (a), we saw that here the value of the $F=$ statistic is 259.27. From Table E, referring to the column labeled by 1 and the row labeled by 17, we see that the P-value is below 0.001.

d) $t^2 = (16.1)^2 = 259.21$. The value of F is 259.27. These agree, up to round-off error.

e) From our ANOVA table, we find

$$r^2 = SSM/SST = 3.7619/4.0085 = 0.938$$

Exercise 10.29

a) Here $r = 0.19$ and $n = 713$. Thus

$$t = \frac{r\sqrt{n-2}}{\sqrt{1-r^2}} = \frac{-0.19\sqrt{713-2}}{\sqrt{1-(-0.19)^2}} = \frac{-0.19(26.66)}{\sqrt{0.9639}} = -5.16$$

b) The $t=$ statistic in (a) has a $t(n - 2) = t(711)$ distribution under H_0. Using Table D to estimate the P-value, we find that the P-value is less than 0.001 (we must double the tail probability in Table D since we have a two-sided hypothesis). We would conclude that there is strong evidence that population correlation ρ is different from 0. Simply stating that the null hypothesis is rejected does not tell us anything about how different ρ is from 0, or whether it is larger or smaller than 0.

CHAPTER 11

MULTIPLE REGRESSION

OVERVIEW

Multiple linear regression extends the techniques of simple linear regression to situations involving $p > 1$ explanatory variables $x_1, x_2,\ldots, x_p$. The statistical model for multiple linear regression is

$$y_i = \beta_0 + \beta_1 x_{i1} + \beta_2 x_{i2} + \ldots + \beta_p x_{ip} + \varepsilon_i$$

where $i = 1, 2,\ldots, n$.. The deviations ε_i are assumed to be independent and normally distributed with mean 0 and standard deviation σ. The **parameters** of the model are $\beta_0, \beta_1, \beta_2,\ldots,\beta_p$, and σ. The β's are estimated by $b_0, b_1, b_2, \ldots, b_p$ by the **principle of least-squares**. σ is estimated by

$$s = \sqrt{\text{MSE}} = \frac{\sum e_i^2}{n - p - 1}$$

where the e_i are the **residuals**

$$e_i = y_i - \hat{y}_i$$

and

$$\hat{y}_i = b_0 + b_1 x_{i1} + b_2 x_{i2} + \ldots + b_p x_{ip}$$

In practice, the b's and s are calculated using statistical software.

A **level C confidence interval** for β_j is

$$b_j \pm t^* \mathrm{SE}_{b_j}$$

where t^* is the upper $(1-C)/2$ critical value for the $t(n - p - 1)$ distribution. SE_{b_j} is the standard error of b_j and in practice is computed using statistical software.

The **test of the hypothesis H_0: $\beta_j = 0$** is based on the t **statistic**

$$t = \frac{b_j}{\mathrm{SE}_{b_j}}$$

with P-values computed from the $t(n - p - 1)$ distribution. In practice statistical software is used to carry out these tests.

In multiple regression, interpretation of these confidence intervals and tests depends on the particular explanatory variables in the multiple regression model. The estimate of β_j represents the effect of the explanatory variable x_j when it is added to a model already containing the other explanatory variables.

The test of H_0: $\beta_j = 0$ tells us if the improvement in the ability of our model to predict the response y by adding x_j to a model already containing the other explanatory variables is statistically significant. It does not tell us if x_j would be useful for predicting the response in multiple regression models with a different collection of explanatory variables.

The **ANOVA table** for a multiple regression is analogous to that in simple linear regression. It gives the sum of squares SSM for the model, the sum of squares SSE for error, the total sum of squares SST for all sources of variation, the degrees of freedom DFM, DFE, and DFT for these sums of squares, the mean squares MSM for the model, and the mean squares MSE for error. The sums of squares are

$$\mathrm{SSM} = \sum (\hat{y}_i - \bar{y})^2$$

$$\mathrm{SSE} = \sum (y_i - \hat{y}_i)^2$$

$$\mathrm{SST} = \sum (y_i - \bar{y})^2$$

and the degrees of freedom are

$$\mathrm{DFM} = p$$
$$\mathrm{DFE} = n - p - 1$$
$$\mathrm{DFT} = n - 1.$$

The mean sums of squares MS are defined by the relation

In practice, these quantities are computed using statistical software. The results are often summarized in an ANOVA table, which usually has a form like the following.

Source	Degrees of freedom	Sum of squares	Mean square	F
Model	DFM	SSM	SSM/DFM	MSM/MSE
Error	DFE	SSE	SSE/DFE	
Total	DFT	SST		

The **ANOVA F=statistic** is the ratio MSM/MSE and is used to test

$$H_0: \beta_1 = \beta_2 = \ldots = \beta_p = 0$$

Under H_0, this statistic has an $F(p, n - p - 1)$ distribution, which can be used to compute P-values using Table E. Notice that evidence against H_0 only tells us that at least one of the β_j differs from 0, but not which one. Deciding which $\beta_j \neq 0$ requires further analysis using the procedures for confidence intervals or hypothesis tests for the individual β_j mentioned above, keeping in mind the difficulties of interpreting these individual inferences.

The **squared multiple correlation** can be written as

$$R^2 = \text{SSM/SST}$$

and is interpreted as the proportion of the variability in the response variable y that is explained by the explanatory variables $x_1, x_2,\ldots, x_p$ in the multiple regression.

SAMPLE PROBLEMS

GUIDED SOLUTIONS

Exercise 11.32

KEY CONCEPTS - simple linear regression, normal quantile plots, residual plots

a) Summarize the results of your analysis by filling in the tables below. We recommend that the analysis be carried out using statistical software.

Source	Degrees of freedom	Sum of squares	Mean square	F
Model				
Error				
Total				

The regression equation is

$$\text{Corn} = \underline{\hspace{1cm}} + \underline{\hspace{1cm}} \text{Year}$$

Variable	Parameter estimate	Std. error	t-ratio	P-value
Constant				
Year	1			

$$s = \sqrt{\text{MSE}} = \underline{\hspace{1.5cm}} \qquad R^2 = \underline{\hspace{1.5cm}}$$

What do you conclude about the significance test for the slope?

b) Use statistical software to make a normal quantile plot. Is there anything in the plot to suggest that the residuals are not normal?

c) Use statistical software to make a plot of the residuals versus soybean yield. Does the plot indicate that soybean yield might be useful in a multiple linear regression model with year to predict corn yield?

Exercise 11.33

KEY CONCEPTS - simple linear regression, normal quantile plots, residual plots

a) Summarize the results of your analysis by filling in the tables below. We recommend that the analysis be carried out using statistical software.

Source	Degrees of freedom	Sum of squares	Mean square	F
Model				
Error				
Total				

The regression equation is

$$\text{Corn} = \underline{\hspace{1cm}} + \underline{\hspace{1cm}}\text{Soybeans}$$

Variable	Parameter estimate	Std. error	t-ratio	P-value
Constant				
Soybeans	1			

$$s = \sqrt{\text{MSE}} = \underline{\hspace{1.5cm}} \qquad R^2 = \underline{\hspace{1.5cm}}$$

What do you conclude about the significance test for the slope?

b) Use statistical software to make a normal quantile plot. Is there anything in the plot to suggest that the residuals are not normal?

c) Use statistical software to make a plot of the residuals versus year. Does the plot indicate that year might be useful in a multiple linear regression model with soybean yield to predict corn yield?

Exercise 11.34

KEY CONCEPTS - multiple linear regression, R^2, interpretation of the regression coefficients in a multiple linear regression, plots of residuals

a) Summarize the results of your analysis by filling in the ANOVA table below. We recommend that the analysis be carried out using statistical software.

Source	Degrees of freedom	Sum of squares	Mean square	F
Model				
Error				
Total				

$$s = \sqrt{MSE} = \underline{\hspace{1.5cm}} \qquad R^2 = \underline{\hspace{1.5cm}}$$

The null and alternative hypotheses tested by the ANOVA F test are

H_0:

H_a:

The P-value of your test of H_0 is _____ (you may be able to read this directly from the statistical software package you use).

What do you conclude?

b) Remember, $R^2 = SSM/SST$ is interpreted as the proportion of the variability in the response variable y that is explained by the explanatory variables $x_1, x_2, ..., x_p$ in the multiple regression. What is the value of R^2 here? (You may be able to read this directly from the statistical software package you use.) Compare this with the values found in the previous two exercises (for the values of R^2 in these exercises, see their complete solutions, which are given below).

c) From the results you obtained with your statistical software, fill in the blanks below.

The regression equation is

$$\text{Corn} = \underline{\hspace{2cm}} + \underline{\hspace{2cm}}\text{Year} + \underline{\hspace{2cm}}\text{Soybeans}$$

Why do the coefficients of year and soybeans differ from those in the previous two exercises? To answer this question, recall that, in general, for the multiple linear regression model

$$y_i = \beta_0 + \beta_1 x_{i1} + \beta_2 x_{i2} + \dots + \beta_p x_{ip} + \varepsilon_i$$

the estimate of β_j represents the effect of the explanatory variable x_j when it is added to a model already containing the other explanatory variables.

d) Summarize the results of the significance tests for the regression coefficients for year and soybean yield by completing the table below. You should be able to find the necessary information from your statistical software.

Variable	Parameter estimate	Std. error	t-ratio	P-value
Constant				
Year				
Soybeans				

What do you conclude?

e) Recall that a 95% confidence interval for β_j is

$$b_j \pm t^* \text{SE}_{b_j}$$

where t^* is the upper 0.025 critical value for the $t(n - p - 1) = t(37)$ distribution, which in this case (see Table D in the text) is approximately 2.03. SE_{b_j} is the standard error of b_j and can be found as an entry in the table you completed in (d). Thus, our 95% confidence intervals here are

For year: $b_1 \pm 2.03\,\text{SE}_{b_1} = $

For soybeans: $b_2 \pm 2.03 \, SE_{b_2} =$

f) Use statistical software to make the residual plots. Do you see any patterns? (How do these plots compare with those in the previous two exercises?) What do you conclude?

COMPLETE SOLUTIONS

Exercise 11.32

a) A summary of the results of a simple linear regression analysis are given below. The numbers were obtained from Minitab but have been reformatted.

Source	Degrees of freedom	Sum of squares	Mean square	F
Model	1	18650	18650	192.04
Error	38	3690	97	
Total	39	22340		

The regression equation is

$$\text{Corn} = -3606 + 1.87 \, \text{Year}$$

Variable	Parameter estimate	Std. error	t-ratio	P-value
Constant	−3605.6	266.8	−13.51	0.000
Year	1.8706	0.1350	13.86	0.000

$s = \sqrt{\text{MSE}} = 9.855$ $R^2 = 83.5\%$

From these results, we see that the *P*-value for the significance test for the slope is very small (below 0.001). This can be seen in the entry for the *P*-value of the parameter estimate for Year (which is the slope). Equivalently, one can use the ANOVA *F* statistic to and Table E to determine this *P*-value.

b) A normal quantile plot (from Minitab) of the residuals is given below.

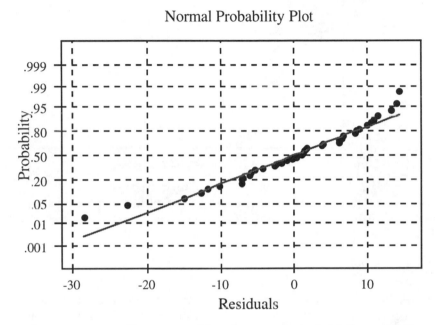

These points appear to lie reasonably close to a straight line and there are no striking features which would suggest that the data are not normal.

c) A plot of the residuals versus soybean yield is given below.

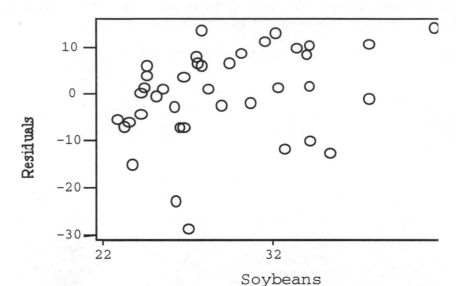

Soybeans

There is some evidence of an upward trend in the plot, suggesting that soybean yield might be useful in a multiple linear regression model with year to predict corn yield.

Exercise 11.33

a) A summary of the results of a simple linear regression analysis are given below. The numbers were obtained from Minitab but have been reformatted.

Source	Degrees of freedom	Sum of squares	Mean square	F
Model	1	19570	19570	268.44
Error	38	2770	73	
Total	39	22340		

The regression equation is

$$\text{Corn} = -47.2 + 4.80 \text{ Soybeans}$$

Variable	Parameter estimate	Std. error	t-ratio	P-value
Constant	−47.199	8.578	-5.50	0.000
Soybeans	4.8031	0.2932	16.38	0.000

$s = \sqrt{\text{MSE}} = 8.538$ $R^2 = 87.6\%$

From these results, we see that the P-value for the significance test for the slope is very small (below 0.001). This can be seen in the entry for the P-value of the parameter estimate for Soybeans (which is the slope). Equivalently, one can use the ANOVA F statistic to and Table E to determine this P-value.

b) A normal quantile plot (from Minitab) of the residuals is given below.

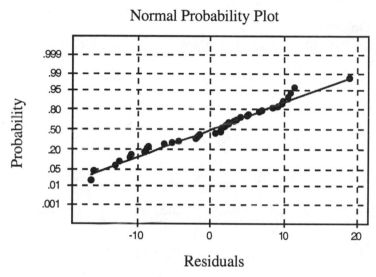

Normal Probability Plot

These points appear to lie reasonably close to a straight line and there are no striking features which would suggest that the data are not normal.

c) A plot of the residuals versus year is given below.

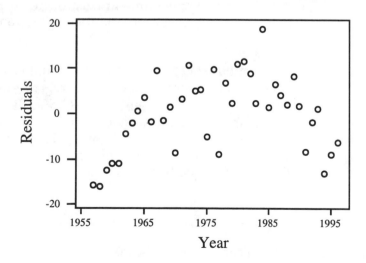

This plot has a curved trend, rising until around 1980 then falling. This suggests that the square of year (rather than year) might be useful in a multiple linear regression model with soybean yield to predict corn yield.

Exercise 11.34

a) The ANOVA table is

Source	Degrees of freedom	Sum of Squares	Mean Square	F
Model	2	20446	10223	199.73
Error	37	1894	51	
Total	39	22340		

$s = \sqrt{MSE} = 7.154$ $R^2 = 91.5\%$

The null and alternative hypotheses tested by the ANOVA F test are

$$H_0: \beta_1 = \beta_2 = 0$$

$$H_a: \text{at least one of } \beta_1 \text{ and } \beta_2 \text{ is not } 0.$$

where β_1 is the regression coefficient of year and β_2 the regression coefficient of soybean yield in our multiple linear regression model. The ANOVA F statistic has an $F(2, n - 3) = F(2, 37)$ distribution. If your statistical software does not give the P-value, you can approximate it from Table E in which case one finds the P-value is less than 0.001. We conclude that at least one of the two regression coefficients is different from 0 in the population regression equation.

b) In our multiple regression model, from our statistical software we find (see (a)) $R^2 = 91.5\%$. This is a little higher than for the simple linear regression model with year as the explanatory variable, where we found $R^2 = 83.5\%$. It is also slightly higher than for the simple linear regression model with soybean yield as the explanatory variable where we found $R^2 = 87.6\%$. Thus the multiple linear regression model explains a little bit higher percentage of the variability in the corn yield than either of the simple linear regression models of the previous two exercises.

c) From our statistical software (Minitab in our case) we get the fitted model

Corn = - 1638 + 0.831 Year + 2.98 Soybeans.

The coefficients differ because they have a different meaning here than in the simple linear regression models. The regression coefficient for year tells us the effect of year as an explanatory variable for corn yield in a model that includes the explanatory variable soybean yield. Soybean yield may capture some of the effect of year, producing a different value for the regression coefficient than we found in the simple linear regression model. Soybean yield was not present as an explanatory variable in the simple linear regression with year as an explanatory variable. Likewise, the regression coefficient for soybean yield tells us the effect of soybean yield as an explanatory variable for corn yield in a model that includes the explanatory variable year. Year may capture some of the effect of soybean yield, producing a different value for the regression coefficient than we found in the simple linear regression model. Year was not present as an explanatory variable in the simple linear regression with soybean yield as an explanatory variable.

d) From our statistical software (Minitab in this case), we get the significance test results for the regression coefficients for year and soybean yield to be

Variable	Parameter estimate	Std. error	t-ratio	P-value
Constant	−1637.9	384.5	−4.26	0.000
Year	0.8314	0.2009	4.14	0.000
Soybeans	2.9837	0.5036	5.92	0.000

We see that there is strong (P-values below 0.001) statistical evidence that both regression coefficients differ from 0 in the multiple linear regression model.

e) Our 95% confidence intervals are

For year: $b_1 \pm 2.03\,\text{SE}_{b_1} = 0.8314 \pm 2.03(0.2009) = 0.8314 \pm 0.4078$

For soybeans: $b_2 \pm 2.03\,\text{SE}_{b_2} = 2.9837 \pm 2.03(0.5036) = 2.9837 \pm 1.0223$

f) Plots of the residuals versus year and versus soybean yield are given below.

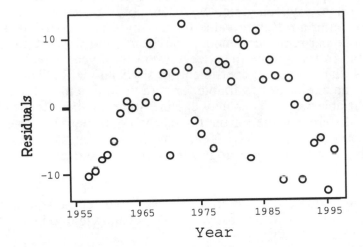

This plot has a curved trend, rising until around 1980 and then falling. This suggests that the square of year might be useful in a multiple linear regression model with soybean yield and year to predict corn yield.

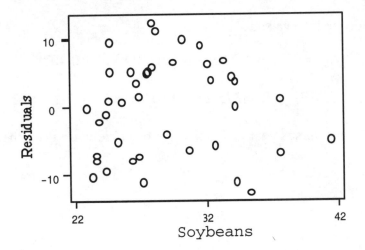

This plot shows no obvious trend, although there may be the merest hint of a slightly curved pattern (residuals higher in the middle than at the ends). This may be due to the pattern observed in the residual plot versus year and the fact that year and soybeans are probably correlated.

CHAPTER 12

ONE-WAY ANALYSIS OF VARIANCE

OVERVIEW

The **one-way analysis of variance** is a generalization of the two sample *t* procedures. It allows comparison of more than two populations based on independent SRSs from each. As in the pooled two-sample *t* procedures, the populations are assumed to be normal with possibly different means, but with a common standard deviation. In the one-way analysis of variance we are interested in making formal inferences about the population means.

The simplest graphical procedure to compare the populations is to give side-by-side boxplots (see Chapter 1). Normal quantile plots can be used to check for extreme deviations from normality or outliers. The summary statistics required for the analysis of variance calculations are the means and standard deviations of each of the samples. An informal procedure to check the assumption of equal variances is to make sure that the ratio of the largest to the smallest standard deviation is less than 2. If the standard deviations satisfy this criterion and the normal quantile plots seem satisfactory, then the one-way ANOVA is an appropriate analysis.

The **F statistic** computed in the ANOVA table can be used to test the **null hypothesis** that the population means are all equal. The **alternative hypothesis** is that at least two of the population means are not equal. Rejection of the null hypothesis does not provide any information as to which of the population means are different.

If the researcher has specific questions about the population means before examining the data, these questions can often be expressed in terms of **contrasts**. Tests and confidence intervals about contrasts provide answers to these questions and allow the researcher to say more about which population means are different and what the sizes of these differences are.

When there are no specific questions before examining the data, **multiple comparisons** are often used to follow up rejection of the null hypothesis in a one-way analysis of variance. These multiple comparisons are designed to determine which pairs of population means are different and to give confidence intervals for the differences.

SAMPLE PROBLEMS

GUIDED SOLUTIONS

Exercise 12.23

KEY CONCEPTS - relationships among entries in the ANOVA table

a) The key relationships among the entries in the ANOVA table required to complete the table are

- the mean squares are equal to the sums of squares divided by their degrees of freedom
- the sum of squares for groups and the sum of squares for error add to the total sum of squares
- the degrees of freedom for groups and the degrees of freedom for error add to the total degrees of freedom (which is $n - 1$)
- the F statistic is the mean square for groups divided by the mean square for error

Use these relationships to complete the ANOVA table below.

Source	Degrees of freedom	Sum of squares	Mean square	F
Groups	3		158.96	
Error	32		62.81	
Total				

b) In the one-way ANOVA table, the F statistic is always used to test the null hypothesis that the means are equal and the alternative hypotheses that they are not equal. There aren't one- and two-sided tests as there were in some of the previous chapters. When writing down these hypotheses, you should try and phrase them in language or notation related to the particular data set.

c) Under the null hypothesis the F statistic has an $F(I - 1, N - I)$ distribution. The degrees of freedom for the F always agree with the degrees of freedom for groups and error, respectively in the ANOVA table. This exercise can be done by identifying the values of I and N in this data set, or by reading the appropriate degrees of freedom from the ANOVA table. You then need to go to Table E and see which F critical values in the table correspond to the value of the F statistic computed in this exercise.

d) Which entry in the ANOVA table corresponds to the estimate of the within-group variance?

Exercise 12.31

KEY CONCEPTS - comparisons among the means in the one-way ANOVA, contrasts

We reproduce the data for the four groups below.

Group	n	$\bar{x}$	s
Treatment (T)	10	51.90	6.42
Control (C)	5	57.40	10.46
Joggers (J)	11	49.73	6.27
Sedentary (S)	10	58.20	9.49

a) Contrasts are comparisons among the means designed to answer specific questions. These questions are posed before the data are collected. There are four means in this problem, μ_T, μ_C, μ_J and μ_S. Give each of the contrasts as a combination of some of these means. Remember that the sum of the coefficients of the means will add to zero for a contrast. The specific question being asked can also be stated in terms of hypotheses about the contrast.

1) Is T better than C? The contrast is $\psi_1 = \mu_C - \mu_T$. The hypotheses are $H_0: \psi_1 = 0$ and $H_a: \psi_1 > 0$. As usual, the alternative is the effect we are trying to find and $\psi_1 > 0$ corresponds to $\mu_C > \mu_T$ or T better than C, since lower depression scores are "better".

2) Is T better than the average of C and S? The contrast is

$\psi_2 =$

The hypotheses are H$_0$:   and H$_a$:

          3) Is J better than the average of the other three groups? The contrast is

$\psi_3 =$

The hypotheses are H$_0$: ·   and H$_a$:

b) The estimate of a contrast or combination of population means $\psi = \sum a_i\mu_i$ is given by the same combination of the sample means. The sample contrast or estimator of ψ is $c = \sum a_i\bar{x}_i$. The standard error of the contrast is given by

$$SE_c = s_p\sqrt{\sum \frac{a_i^2}{n_i}} \, ,$$

and a test of the null hypothesis that the contrast is 0 uses the t statistic,

$t = \dfrac{c}{SE_c}$. The first thing that needs to be done when applying these results in a particular setting is to identify the a_i in the contrast. In the three contrasts from (a) of this exercise, the first is a simple difference of means, while the second and third are more complicated. The answer below applies these general results about contrasts to the second contrast, leaving the the first and third contrast for you to practice on.

          1) Is T better than C?

Sample contrast =

Standard error =

t statistic =   df P-value =

          2) Is T better than the average of C and S? The contrast is

$$\psi_2 = \frac{1}{2}(\mu_C + \mu_S) - \mu_T.$$

Remember the sample contrast is the same combination as the population contrast with the population means replaced by the sample means.

Sample contrast $= \frac{1}{2}(\bar{x}_C + \bar{x}_S) - \bar{x}_T = \frac{1}{2}(57.40 + 58.20) - 51.90 = 5.9$

The standard error of a contrast requires the within-groups estimate of the standard deviation s_p, and the a_i and n_i for the contrast. The within-groups estimate of the standard deviation s_p, corresponds to the square root of the MSE in the ANOVA table. The ANOVA table for this data is given in Exercise 12.23 and $s_p = \sqrt{62.81} = 7.93$.

The a_i for this constrast are 1/2, 1/2, and -1. These a_i correspond to sample sizes n_i of 5, 10 and 10 respectively. We can now apply the general formula for the standard error of a contrast.

$$\text{Standard error} = \text{SE}_c = s_p \sqrt{\sum \frac{a_i^2}{n_i}} = 7.93 \sqrt{\frac{(1/2)^2}{5} + \frac{(1/2)^2}{10} + \frac{(-1)^2}{10}} = 3.3174.$$

$$t \text{ statistic} = \frac{c}{\text{SE}_c} = \frac{5.9}{3.3174} = 1.78$$

Since the alternative is one-sided, the P-value is $P(t > 1.78)$ where the t has the degrees of freedom that are associated with s_p which is $n - I = 32$. Using Table D we see that the P-value is between 0.025 and 0.05, or using statistical software the value is 0.0423.

       3) Is J better than the average of the other three groups?

Sample contrast =

Standard error =

t statistic =   df P-value =

Does the use of contrasts give an adequate summary of the results? How does the information from the contrasts compare with the information contained in simply presenting the ANOVA table and the associated F statistic?

c) Is this observational data or an experiment? Can we conclude that there is a cause and effect relationship or is there only an association?

Exercise 12.35

KEY CONCEPTS - Bonferroni multiple comparisons

Multiple comparison methods are designed to determine which pairs of means are different. These methods are used when the null hypothesis is rejected, and we didn't have specific questions about the means in advance of the analysis. Multiple comparison procedures are performed by computing t statistics for all pairs of means using the formula

$$t_{ij} = \frac{\bar{x}_i - \bar{x}_j}{s_p \sqrt{\dfrac{1}{n_i} + \dfrac{1}{n_j}}}$$

This is the same t statistic that would be used when studying the contrast $\psi = \mu_i - \mu_j$. We declare the population means μ_i and μ_j different whenever

$$| t_{ij} | \geq t^{**},$$

where the value of t^{**} depends on the multiple comparison procedure being applied. In this exercise, the Bonferroni multiple comparison procedure is being used and the value of t^{**} is given in the exercise as $t^{**} = 2.53$. The simulaneous confidence intervals which tell you about the the *sizes* of the differences between the means are found using the formula

$$\bar{x}_i - \bar{x}_j \pm t^{**} s_p \sqrt{\frac{1}{n_i} + \frac{1}{n_j}}.$$

When doing the calculations, it is often simplest to summarize them in a table like the one below. The details for the first pair of means (T, C) are worked out for you. Note that the quantity $s_p \sqrt{\dfrac{1}{n_i} + \dfrac{1}{n_j}}$ which appears in the calculations of both the t statistic and the confidence interval, has the same value for any two groups with the same pair of sample sizes such as (T, C) and (C, S). Use of this fact will save some time in the calculations.

For the pair (T, C),
$$\bar{x}_T - \bar{x}_C = 51.90 - 57.40 = -5.5,$$

$$s_p \sqrt{\frac{1}{n_T} + \frac{1}{n_C}} = 7.93 \sqrt{\frac{1}{10} + \frac{1}{5}} = 4.3434$$

The t statistic is

$$t_{TC} = \frac{\bar{x}_T - \bar{x}_C}{s_p\sqrt{\dfrac{1}{n_T} + \dfrac{1}{n_C}}} = \frac{-5.5}{4.3434} = \text{-}1.27$$

The confidence interval for $\mu_T - \mu_C$ is

$$\bar{x}_T - \bar{x}_C \pm t^{**} s_p\sqrt{\frac{1}{n_T} + \frac{1}{n_C}} = \text{-}5.5 \pm 2.53 \times 4.3434 = \text{-}5.5 \pm 10.99$$

Pair of means	$\bar{x}_i - \bar{x}_j$	$s_p\sqrt{\dfrac{1}{n_i} + \dfrac{1}{n_j}}$	t_{ij}	Confidence interval
(T, C)	--5.50	4.3434	-1.27	-17.06 $\pm$ 10.99
(T, J)				
(T, S)				
(C, J)				
(C, S)				
(J, S)				

How would you interpret these results?

COMPLETE SOLUTIONS

Exercise 12.23

a) The mean square for groups is the group sum of squares divided by its degrees of freedom. We are given the mean square for groups, so the group sum of squares is the mean square multiplied by its degrees of freedom, 3 x 158.96 = 476.88. Similarly, the sum of squares for error is 32 x 62.81 = 2009.92. The other entries follow the rules given in the guided solution.

Source	Degrees of freedom	Sum of squares	Mean square	F
Groups	3	476.88	158.96	2.53
Error	32	2009.92	62.81	
Total	35	2486.80		

b) The null hypothesis is that the mean depression scores for the four populations from which the samples are assumed to be drawn are equal. The alternative hypothesis is that the means are different, but does not specify what these differences might be. Using the notation, we have the hypotheses

H_0: $\mu_T = \mu_C = \mu_J = \mu_S$ and H_a: μ_T, μ_C, μ_J, and μ_S are not all equal.

c) Under the null hypothesis the F statistic has an $F(I - 1, N - I) = F(3, 32)$ distribution since the number of groups $I = 4$ and the number of subjects $N = 36$. Referring to Table E, we find that the F critical values for tail probabilities 0.10 and 0.05 are 2.28 and 2.92 respectively, where we have gone to the $F(3, 30)$ distribution to be conservative, since 32 degrees of freedom is not in the table. The P-value is between 0.05 and 0.10. Computer software using the degrees of freedom 3 and 32 gives the P-value as 0.0747. There is some evidence of a difference in the depression scores between the four groups, but the evidence is not that strong.

d) The mean square for error is the estimate of the within-group variance, so $s_p^2 = 62.81$. The square root is $s_p = \sqrt{62.81} = 7.93$.

Exercise 12.31

a) The first contrast is given in the guided solution.

   2) Is T better than the average of C and S? The contrast is

$$\psi_2 = \frac{1}{2}(\mu_C + \mu_S) - \mu_T.$$

The hypotheses are H_0: $\psi_2 = 0$ and H_a: $\psi_2 > 0$. The alternative consists of parameter values for which the average of C and S is greater than T, or since lower scores are better, that T is better than the average of C and S.

   3) Is J better than the average of the other three groups? The contrast is

$$\psi_3 = \frac{1}{3}(\mu_C + \mu_S + \mu_T) - \mu_J. \text{ The hypotheses are } H_0: \psi_3 = 0 \text{ and } H_a: \psi_3 > 0.$$

b)
   1) Is T better than C? $\psi_1 = \mu_C - \mu_T$.

Sample contrast = $\bar{x}_C - \bar{x}_T = 57.40 - 51.90 = 5.5$

$$\text{Standard error} = \text{SE}_c = s_p \sqrt{\sum \frac{a_i^2}{n_i}} = 7.93 \sqrt{\frac{1^2}{5} + \frac{(-1)^2}{10}} = 4.34.$$

$$t \text{ statistic} = \frac{c}{\text{SE}_c} = \frac{5.5}{4.34} = 1.27 \qquad df = 32, \ P\text{-value} = 0.1066 \text{ (using software)}$$

2) See guided solution.

3) The contrast is $\psi_3 = \dfrac{1}{3}\left(\mu_C + \mu_S + \mu_T\right) - \mu_J$.

$$\text{Sample contrast} \qquad = \frac{1}{3}\left(\bar{x}_C + \bar{x}_S + \bar{x}_T\right) - \bar{x}_J$$

$$= \frac{1}{3}(57.40 + 58.20 + 51.90) - 49.73 = 6.10$$

$$\text{Standard error} = \text{SE}_c = s_p \sqrt{\sum \frac{a_i^2}{n_i}} = 7.93 \sqrt{\frac{(1/3)^2}{5} + \frac{(1/3)^2}{10} + \frac{(1/3)^2}{10} + \frac{(-1)^2}{11}}$$

$$= 2.9175.$$

$$t \text{ statistic} = \frac{c}{\text{SE}_c} = \frac{6.1}{2.9175} = 2.09 \quad df = 32, \ P\text{-value} = 0.0223 \text{ (using software)}$$

The use of contrasts provides specific answers as to the nature of some of the differences between the four groups. The ANOVA table and associated F statistic just tell us whether there is evidence of a difference between the four groups, but we can't say more. The use of contrasts allows us to identify where the differences are and answer specific questions about comparisons between the groups which are of interest.

c) This is observational data. The experimenter did not assign the male subjects to be in an exercise group, jog, or have a sedentary lifestyle. There is a possibility the groups differed on other (confounding) variables which might be related to depression. The description of the data in Exercise 12.22 indicates that the experimenter tried to select subjects similar in age and other characteristics. While this is desirable and may eliminate certain potential confounding variables such as age, we can't be sure that the experimenter has balanced the groups on all potential confounding variables. Randomization of subjects to the treatment groups is still the best way to do this.

Exercise 12.35

Below is the completed table for all pairwise comparison of using the Bonferroni method.

Pair of means	$\bar{x}_i - \bar{x}_j$	$s_p\sqrt{\dfrac{1}{n_i}+\dfrac{1}{n_j}}$	t_{ij}	Confidence interval
(T, C)	-5.50	4.3434	-1.27	-5.50 ± 10.99
(T, J)	2.17	3.1649	0.63	2.17 ± 8.77
(T, S)	-6.30	3.5464	-1.77	-6.30 ± 8.97
(C, J)	7.67	4.2771	1.79	7.67 ± 10.82
(C, S)	-0.80	4.3434	-0.18	-0.80 ± 10.99
(J, S)	-8.47	3.4649	-2.44	-8.47 ± 8.77

Using an error probability of 0.05 we would conclude that there is no evidence of a difference between any pair of means since none of these t_{ij} values exceeds 2.53 in absolute value. This is the same conclusion we reached when using the *F* statistic in the ANOVA table.

The interpretation of 95% *simultaneous* confidence intervals is that, in the long run, we are requiring that 95% of the time *all six* intervals contain the true differences in means, rather than only requiring this property to hold for a single interval. In order to accomplish this, the intervals need to be much longer than an individual confidence interval. Similarly, when doing the hypothesis tests, the error probability is controlling the probability of *any* false rejection among the six tests, so the *t*** value is much larger than it would be for an individual test. The multiple comparison approach can be quite conservative and it is not surprising that some of the contrasts in Exercise 12.31 were significantly different from zero, while based on the multiple comparison procedure we would conclude all pairs of means are equal. To overcome the conservative nature of multiple comparisons, some researchers choose an error probability that is a little higher than you might ordinarily use when performing a single test, such as 0.10 for example.

Finally, remember that not rejecting the null hypothesis is not the same as demonstrating that it is true. It is still a good idea to look at the confidence intervals and the estimates of the differences, to see if any of the comparisons are deserving of further study, possibly with a larger sample size.

CHAPTER 13

TWO-WAY ANALYSIS
OF VARIANCE

OVERVIEW

The two-way analysis of variance is designed to compare the means of populations that are classified according to two factors. As with the one-way ANOVA, the populations are assumed to be normal with possibly different means and the same standard deviation. The observations are independent SRSs drawn from each population.

The preliminary data summary should include examination of means and standard deviations, and normal quantile plots. Typically, the means are summarized in a two-way table with the rows corresponding to the level of one factor and the columns corresponding to the levels of the second factor. The **marginal means** are computed by taking averages of these cell means across the rows and columns. These means are typically plotted so that the **main effects** of each factor, as well as their **interaction** can be examined.

In the two-way ANOVA table, the **model** variation is broken down into parts due to each of the main effects and a third part due to the interaction. In addition, the ANOVA table organizes the calculations required to compute F statistics and P-values to test hypotheses about these main effects and their interaction. The within-group variance is estimated by pooling the standard deviations from the cell and corresponds to the mean square for error in the ANOVA table.

GUIDED SOLUTIONS

Exercise 13.7

KEY CONCEPTS - plotting group means, main effects, interactions

a) Complete the plot of the mean GITH for these diets on the axes provided below. You should plot the two means when Chromium is Low above the symbol L, and the two means when Chromium is Normal above the symbol N. Then, for each Eat group, connect the points for the two Chromium means.

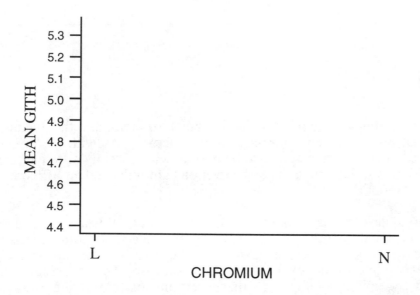

b) In Exercise 13.17, we will see which effects in the plot are real and which are due to chance variation. For now, try to identify some of the main features of the plot. Do the lines seem to be approximately parallel? Is the effect of Chromium similar for both levels of Eat? Are the lines for the two levels of Eat far apart? Is there much change in the mean GITH when going from Low to Normal levels of Chromium? Think about what the answers to these different questions are telling you about the factors Eat and Chromium.

c) In the table below, fill in the marginal means.

Chromium	Eat	
	M	R
L	4.545	5.175
N	4.425	5.317

Compare the difference between the M and R diets for each level of Chromium. How does this comparison show up in your plot?

Exercise 13.11

KEY CONCEPTS - computing group means and marginal means, two-way ANOVA

a) In the table below, fill in the sample sizes, means, and standard deviations for the six species-size groups. You should try to do the computations using computer software.

group	N	Mean	StDev
Aspen, S1			
Birch, S1			
Maple, S1			
Aspen, S2			
Birch, S2			
Maple, S2			

In the table below, fill in the means for the different species-size groups and the marginal means for size and species. This type of table is produced easily using most computer software packages.

	Aspen	Birch	Maple	ALL
S1				
S2				
ALL				

b) Plot the means of the six groups computed in (a) on the axes given. The species is on the x axis, and the modulus of elasticity is on the y axis. For each size, connect the three means corresponding to the different species.

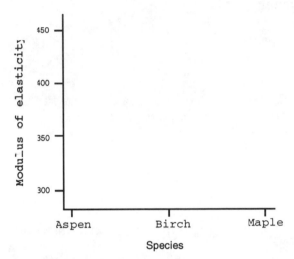

Describe any patterns you see. Focus on the differences among species, sizes, and any interaction between them.

c) Use your computer software to generate an ANOVA table similar to the one below.

```
Analysis of Variance for TENSION MODULUS OF ELASTICITY

Source          DF        SS         MS        F       P
SIZE             1       3308       3308     0.27   0.613
SPECIES          2      14511       7256     0.59   0.569
SIZE*SPECIES     2      41707      20854     1.70   0.224
Error           12     147138      12262
Total           17     206664
```

Are the results of the significance tests in agreement with the impressions that you formed from the graphs in (b)? If the results are not what you expected, try to come up with a reason for the discrepancies between the ANOVA table and the plots.

Exercise 13.17

KEY CONCEPTS - two-way ANOVA table, drawing conclusions

a) Complete the ANOVA table below. The first step is to fill in the degrees of freedom. Then the mean squares can be obtained easily from the sums of squares and the F's as the ratio of the appropriate mean squares.

Source	Degrees of freedom	Sum of squares	Mean square	F
A(Chromium)			0.00121	
B(Eat)			5.79121	
AB			0.17161	
Error			0.03002	
Total				

b) The degrees of freedom for an F statistic are the degrees of freedom for the mean square in the numerator and the degrees of freedom for the mean square in the denominator, respectively. When you have determined the value of the F statistic and its degrees of freedom, you need to compare the F-value to the critical values provided in Table E. When using Table E the best that can be done is to give two values that the P-value must fall between. What is your conclusion?

c) Carry out the tests for the main effects of Chromium and Eat. What do you conclude?

d) The within-group variance corresponds to which entry in the ANOVA table?

e) Summarize the results of the experiment, using what you have learned in this exercise and in Exercise 13.7.

COMPLETE SOLUTIONS

Exercise 13.7

a)

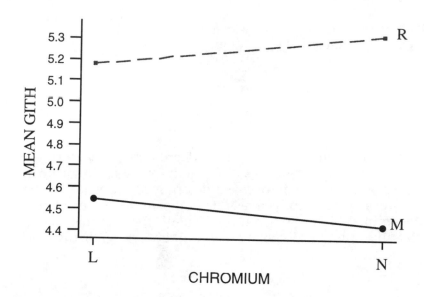

Interaction Plot - Means for GITH

b) The plot suggests that there may be an interaction. The effect of Chromium when going from Low to Normal levels is to decrease mean GITH when the rats could eat as much as they wanted (M) and to increase the mean GITH when the total amount the rats could eat was restricted (R). Without a formal hypothesis test, we don't know whether this apparent interaction is due to chance variation or the effect is real. The effect of Chromium appears to be small compared to the effect of Eat. The two lines are quite far apart, showing a larger effect of Eat. The change in going from the Low to Normal levels of Chromium is much smaller.

c)

	Eat		Mean
Chromium	M	R	
L	4.545	5.175	4.860
N	4.425	5.317	4.871
Mean	4.485	5.246	4.866

At the Low level of Chromium, the difference between M and R is –0.63, and at the Normal level of Chromium, the difference between M and R is –0.892. In the plot, the two means at the Low level of Chromium are closer together than the two means at the Normal level of Chromium. This is reflected in the fact that the two lines are not parallel.

Exercise 13.11

a) The table below gives the sample sizes, means and standard deviations for the six species-size groups.

group	N	Mean	StDev
Aspen, S1	3	387.3	68.7
Birch, S1	3	292.7	121.8
Maple, S1	3	323.3	52.0
Aspen, S2	3	335.7	60.1
Birch, S2	3	455.3	117.7
Maple, S2	3	293.7	183.9

The table below gives the means for the different species-size groups and the marginal means for size and species.

ROWS: SIZE   COLUMNS: SPECIES

	Aspen	Birch	Maple	ALL
S1	387.33	292.67	323.33	334.44
S2	335.67	455.33	293.67	361.56
ALL	361.50	374.00	308.50	348.00

b)

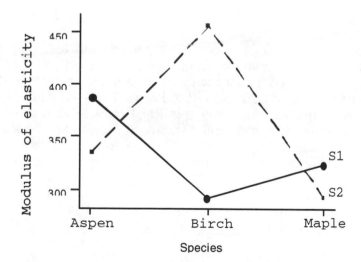

The most striking feature in the plot is the difference in the profiles of mean over the three species for the two sizes of flakes. This type of plot is suggestive of an interaction as the lines are far from parallel. In terms of Species and Size, the species exhibit greater differences than size, as shown in the table of the marginal means.

c) The ANOVA table is reproduced below.

```
Analysis of Variance for TENSION MODULUS OF ELASTICITY

Source              DF          SS          MS          F        P
SIZE                 1        3308        3308       0.27    0.613
SPECIES              2       14511        7256       0.59    0.569
SIZE*SPECIES         2       41707       20854       1.70    0.224
Error               12      147138       12262
Total               17      206664
```

This is an interesting example, in that the effects that seem quite pronounced in the plots are not statistically significant. Examining the mean squares we see that by far the largest mean square is due to interaction as the plot suggests. But it is not statistically significant. The reason is that the MSE is also quite large due to the tremendous variability in the data within each of the six size-species groups. For example, if you examine the 3 observations for the Maple-S2 group, two of these observations are among the smallest in the data, and the third is one of the largest. With this kind of variability, it is difficult to detect group differences, whether they are main effects or interactions.

Due to the large within group variability, if the sizes of the differences seem to be of practical importance, this study should be repeated with a larger sample size to draw more definitive conclusions.

Exercise 13.17

a) The degrees of freedom for the main effects are the number of levels of the factor minus 1. The degrees of freedom for Chromium are 2 - 1 = 1 and for Eat are 2 - 1 = 1. The degrees of freedom for the interaction is the product of the degrees of freedom for the main effects and is 1 x 1 = 1. The total degrees of freedom is the number of subjects minus 1 which is 40 - 1 = 39. The degrees of freedom for error can be gotten by subtraction. The mean squares are the sums of squares divided by their degrees of freedom, and the F for main effects and interaction are gotten by dividing the mean squares for each by the error mean square.

Source	Degrees of freedom	Sum of squares	Mean square	F
A(Chromium)	1	0.00121	0.00121	0.04
B(Eat)	1	5.79121	5.79121	192.89
AB	1	0.17161	0.17161	5.72
Error	36	1.08084	0.03002	
Total	39	7.04487		

b) Under the null hypothesis, the F statistic used to test the null hypothesis of no interaction has an $F(1, 36)$ distribution. The numerical value of the F statistic is 5.72. Referring this to the critical values in Table E, we see the P-value is between 0.025 and 0.010. Since 36 degrees of freedom is not in the table, we need to look at the entries for 30 and 40. Computer software using the $F(1, 36)$ distribution gives the P-value as 0.0221.

c) The main effect of Chromium has an $F = 0.04$. From Table E, the P-value is greater than 0.10. Computer software using the $F(1, 36)$ distribution gives the P-value as 0.8426. The main effect of Eat has an $F = 192.89$. From Table E, the P-value is less than 0.001. Computer software using the $F(1, 36)$ distribution gives the P-value as 0.0000.

d) The within-group variance is the mean square for error, so $s_p^2 = 0.03002$, and

$$s_p = \sqrt{0.03002} = 0.173.$$

e) The interaction between Eat and Chromium is statistically significant, but the effect is relatively small compared to the main effect of Eat. It would be up to the experimenter to determine whether the difference in the effect of going from Low to Normal levels of Chromium for the two levels of Eat is large enough to be of practical interest. By far, the biggest effect is the main effect of Eat. The mean GITH scores are much smaller when the rats could eat as much as they wanted (M) than when the total amount the rats could eat was restricted (R).

CHAPTER 14

NONPARAMETRIC TESTS

SECTION 14.1

OVERVIEW

Many of the statistical procedures described in previous chapters assumed that the samples were drawn from normal populations. **Nonparametric tests** do not require any specific form for the distributions of the populations from which the samples were drawn. Many nonparametric tests are **rank tests**; that is, they are based on the **ranks** of the observations rather than on the observations themselves. When ranking the observations from smallest to largest, tied observations receive the average of their ranks.

The **Wilcoxon rank sum test** compares two distributions. The objective is to determine if one distribution has systematically larger values than the other. The observations are ranked, and the **Wilcoxon rank sum statistic W** is the sum of the ranks of one of the samples. The Wilcoxon rank sum test can be used in place of the **two-sample t test** when samples are small or the populations are far from normal.

Exact P-values require special tables and are produced by some statistical software. However, many statistical software packages give only approximate P-values based on a normal approximation, typically with a continuity correction employed. Many packages also make an adjustment in the normal approximation when there are ties in the ranks.

GUIDED SOLUTIONS

Exercise 14.5

KEY CONCEPTS - ranking data, two-sample problem, Wilcoxon rank sum test

a) Is this a one-sided or two-sided alternative? It is best to state the null and alternative hypotheses in words. You may wish to reread Example 7.17 in the text before you formulate the hypotheses.

H_0:
H_a:

b) Order the observations from smallest to largest in the space provided. Use a different color, or underline those observations in the control group. This will make it easier to determine the ranks assigned to each group.

Now suppose that the first sample is the control group and the second sample is the DDT group. The choice of which sample we call the first sample and which we call the second sample is arbitrary. However, the Wilcoxon rank sum test is the sum of the ranks of the first sample, and the formulas for the mean and variance of W distinguish between the sample sizes for the first and the second samples. What are the ranks of the control observations? Use these ranks to compute the value of W.

$W =$

What are the values of n_1, n_2, and N? Use these to evaluate the mean and standard deviation of W according to the formulas below.

$$\mu_W = \frac{n_1(N+1)}{2} =$$

$$\sigma_W = \sqrt{\frac{n_1 n_2 (N+1)}{12}} =$$

Now use the mean and standard deviation to compute the standardized rank sum statistic

$$z = \frac{W - \mu_W}{\sigma_W} =$$

What kind of values would W have if the alternative were true? Use the normal approximation (with the continuity correction if you have learned it) to find the approximate P-value. If you have access to software or tables to evaluate the exact P-value, compare it with the approximation.

What are your conclusions?

c) How do the results compare with those obtained using the two-sample t test in Example 7.17?

COMPLETE SOLUTIONS

Exercise 14.5

a) H_0: the distribution of nerve activity is the same for both groups
       H_a: the nerve activity is systematically higher in either the DDT group
              or the control group

The alternative is two-sided, because as indicated in Example 7.17, the researchers did not conjecture in advance that the nerve activity would increase in rats fed DDT.

b) The observations are first ordered from smallest to largest. The observations in bold are from the DDT group.

6.642, 8.182, **8.456**, 9.351, 9.686, 11.074, 12.064, **12.207, 16.869, 20.589, 22.429, 25.050**

Suppose the first sample is the control and the second sample is the DDT group. In this case, $n_1 = n_2 = 6$, and N, the sum of the sample sizes is 12. The control observations have ranks 1, 2, 4, 5, 6, 7 and the sum of these ranks is $W = 25$. The values for the mean and variance are

$$\mu_W = \frac{n_1(N+1)}{2} = \frac{6(12+1)}{2} = 39 \text{ and}$$

$$\sigma_W = \sqrt{\frac{n_1 n_2 (N+1)}{12}} = \sqrt{\frac{(6)(6)(12+1)}{12}} = 6.25,$$

and the standardized rank sum statistic W is $z = \dfrac{W - \mu_W}{\sigma_W} = \dfrac{25-39}{6.25} = -2.24.$ We expect W to be either small or large when the alternative hypothesis is true, as DDT could cause the control group to tend to receive either the smaller or larger ranks. The approximate P-value is $2 \times P(Z \le -2.24) = 0.0250$, where we have doubled the probability since the alternative is two-sided.

If we use the continuity correction, we act as if the number 25 occupies the interval from 24.5 to 25.5. This means to compute the P-value, we first calculate the probability that $W \le 25.5$, giving

$$P(W \le 25.5) = P\left(\frac{W - \mu_W}{\sigma_W} \le \frac{25.5 - 39}{6.25}\right) = P(Z \le -2.16) = 0.0154,$$

and then double this giving the P-value $2 \times 0.0154 = 0.0308$. The exact P-value using tables is $2 \times 0.0130 = 0.0260$. Typically the continuity correction improves the approximation, but in this example the approximation is slightly better without the correction. The data give strong evidence that the measures of nerve activity are systematically higher in the DDT group.

c) The results aren't different enough to change the conclusions of the study.

SECTION 14.2

OVERVIEW

The **Wilcoxon signed rank test** is a nonparametric test for matched pairs. It tests the null hypothesis that there is no systematic difference between the observations within a pair against the alternative that one observation tends to be larger.

The test is based on the **Wilcoxon signed rank statistic W^+**, which provides another example of a nonparametric test using ranks. The absolute values of the observations are ranked and the sum of the ranks of the positive (or negative) differences gives the value of W^+. The **matched pairs t test** and the **sign test** are two other alternative tests for this setting.

P-values can be found from special tables of the distribution or a normal approximation to the distribution of W^+. Some software computes the exact P-value and other software uses the normal approximation, typically with a ties correction. Many packages make an adjustment in the normal approximation when there are ties in the ranks.

GUIDED SOLUTIONS

Exercise 14.17

KEY CONCEPTS - matched pairs, Wilcoxon signed rank statistic

First give the null and alternative hypotheses

H_0:
H_a:

To compute the Wilcoxon signed rank statistic, first order the absolute values of the differences and rank them. Due to the large number of ties in this exercise, you need to be careful when computing the ranks. For any tied group of observations, they should each receive the average rank for the group. (Note that the negative observations are in boldfaced type.)

Absolute values	Ranks
1	1.5
1	1.5
3	4.5
3	4.5
3	4.5
3	4.5
4	7.5
4	7.5
7	9
10	10.5
10	10.5
12	12
22	13
31	14

To see how the ranks are computed, the 1's would get ranks 1 and 2, so their average rank is 1.5. The 3's would get ranks 3, 4, 5, and 6, so their average rank is 4.5, and so on. If W^+ is the sum of the ranks of the positive observations, compute the value of W^+.

$W^+ =$

Evaluate the mean and standard deviation of W^+ according to the formulas below.

$$\mu_{W^+} = \frac{n(n+1)}{4}$$

$$\sigma_{W^+} = \sqrt{\frac{n(n+1)(2n+1)}{24}}$$

Now use the mean and standard deviation to compute the standardized rank sum statistic

$$z = \frac{W^+ - \mu_{W^+}}{\sigma_{W^+}} =$$

Do you expect W^+ to be small or large if the alternative is true? Use the normal approximation to find the approximate P-value. (Note: The continuity correction is often used for statistics whose values are whole numbers. This is true for W^+ when there are no tied observations. However, when there are tied observations and W^+ can take on values that are not whole numbers, the continuity correction would not be employed.)

What are your conclusions?

COMPLETE SOLUTIONS

Exercise 14.17

The null and alternative hypotheses are

> H_0: healing rates have the same distribution for the both groups
> H_a: healing rates are systematically lower in the experimental group

The Wilcoxon signed rank statistic is

$$W^+ = 4.5 + 7.5 + 10.5 = 22.5$$

The values for the mean and variance are

$$\mu_{W^+} = \frac{n(n+1)}{4} = \frac{14(14+1)}{4} = 52.5$$
$$\text{and}$$

$$\sigma_{W^+} = \sqrt{\frac{n(n+1)(2n+1)}{24}} = \sqrt{\frac{(14)(15)(28+1)}{24}} = 15.93$$

and the standardized signed rank statistic W^+ is

$$\frac{W^+ - \mu_{W^+}}{\sigma_{W^+}} \leq \frac{22.5 - 52.5}{15.93} = -1.88.$$

If changing the electric field reduces the healing rate, we would expect the differences (experimental – control) to be negative. Thus the ranks of the positive observations should be small and we would expect the value of the statistic W^+ to be small when the alternative hypothesis is true. The approximate P-value is $P(Z \leq -1.88) = 0.0301$.

The output from the Minitab computer package gives a similar result. Many computer packages, including Minitab, include a correction to the standard deviation in the normal approximation to account for the ties in the ranks.

```
Wilcoxon Signed Rank Test

TEST OF MEDIAN = 0.000000 VERSUS MEDIAN L.T. 0.000000

                 N FOR    WILCOXON
          N      TEST    STATISTIC   P-VALUE
Nerve     14      14        22.5      0.032
```

The data show that changing the electric field reduces the healing rate.

SECTION 14.3

OVERVIEW

The **Kruskal-Wallis test** is the nonparametric test for the **one-way analysis of variance** setting. In comparing several populations, it tests the null hypothesis that the distribution of the response variable is the same in all groups and the alternative hypothesis that some groups have distributions of the response variable that are systematically larger than others.

The **Kruskal-Wallis statistic** H compares the average ranks received for the different samples. If the alternative is true, some of these should be larger than others. Computationally, it essentially arises from performing the usual one-way ANOVA to the ranks of the observations rather than the observations themselves.

P-values can be found from special tables of the distribution or a chi-square approximation to the distribution of *H*. When the sample sizes are not too small, the distribution of *H* for comparing *I* populations has approximately a chi-square distribution with $I - 1$ degrees of freedom. Some software computes the exact *P*-value, and other software uses the chi-square approximation, typically with an adjustment in the chi-square approximation when there are ties in the ranks.

GUIDED SOLUTIONS

Exercise 14.21

KEY CONCEPTS - one-way ANOVA, Kruskal-Wallis statistic

a) Think carefully about the differences in the hypotheses being tested.

ANOVA test

H_0:
H_a:

Kruskal-Wallis test

H_0:
H_a:

b) Find the median for each group. Recall that the median, is the middle observation after the observations have been ordered. When the sample sizes are even, the median is the average of the two middle observations.

Nematodes	Median
0	
1000	
5000	
10000	

Using the information in the medians, do the nematodes appear to retard growth?

To compute the Kruskal-Wallis test statistic, the 16 observations are first arranged in increasing order. That step has been carried out below, where we have kept track of the group for each observation. You need to fill in the ranks in the line provided. Remember that there is one tied observation.

```
Growth    3.2     4.6     5.0     5.3     5.4     5.8     7.4
Group   10000    5000    5000   10000    5000   10000    5000
Rank


Growth    7.5     8.2     9.1     9.2    10.8    11.1    11.1
Group   10000    1000       0       0       0    1000    1000
Rank


Growth   11.3    13.5
Group    1000       0
Rank
```

Now fill in the table below which gives the ranks for each of the nematode groups and the sum of ranks for each group.

Nematodes	Ranks	Sum of Ranks
0		
1000		
5000		
10000		

Use the sum of ranks for the four groups to evaluate the Kruskal-Wallis statistic. What are the numerical values of the n_i and N in the formula?

$$H = \frac{12}{N(N+1)} \sum \frac{R_i^2}{n_i} - 3(N+1) \ =$$

The value of H is compared with critical values in Table G for a chi-square distribution with $I-1$ degrees of freedom, where I is the number of groups. What is the P-value, and what do you conclude?

COMPLETE SOLUTIONS

Exercise 14.21

a) The null and alternative hypotheses for the ANOVA test are

H_0: $\mu_0 = \mu_{1000} = \mu_{5000} = \mu_{10000}$
H_a: not all four means are equal

and the null and alternative hypotheses for the Kruskal-Wallis test are

H_0: seedling growths have the same distribution in all groups
H_a: seedling growth is systematically higher in some groups than in others

When the distributions have the same shape, the null hypothesis for the Kruskal-Wallis is that the median growth in all groups is equal, and the alternative hypothesis is that not all four medians are equal.

b) The medians for the nematode groups are given below.   The ordered observations from the first group are 9.1, 9.2, 10.8, 13.5. The median is the average of the two middle observations, $(9.2 + 10.8)/2 = 10.0$.

Nematodes	Median
0	10.00
1000	11.10
5000	5.20
10000	5.55

The medians suggest that nematodes reduce growth rate. There appears not to be a decrease when nematodes are at the level of 1000, but the growth drops off at 5000, and no further reduction is apparent at 10000.

It is important to note that neither the ANOVA test nor the Kruskal-Wallis test are designed specifically for the alternative that nematodes retard growth (this would be the analog of a one-sided alternative). Both procedures test the alternative of *any* differences between the groups.

The computations required for the Kruskal-Wallis test statistic are summarized in the tables below.

Growth	3.2	4.6	5.0	5.3	5.4	5.8	7.4
Group	10000	5000	5000	10000	5000	10000	5000
Rank	1	2	3	4	5	6	7

Growth	7.5	8.2	9.1	9.2	10.8	11.1	11.1
Group	10000	1000	0	0	0	1000	1000
Rank	8	9	10	11	12	13.5	13.5

Growth	11.3	13.5
Group	1000	0
Rank	15	16

Nematodes	Ranks	Sum of Ranks
0	10, 11, 12, 16	49
1000	9, 13.5, 13.5, 15	51
5000	2, 3, 5, 7	17
10000	1, 4, 6, 8	19

$$H = \frac{12}{N(N+1)} \sum \frac{R_i^2}{n_i} - 3(N+1) = \frac{12}{16(16+1)} \left(\frac{49^2}{4} + \frac{51^2}{4} + \frac{17^2}{4} + \frac{19^2}{4} \right) - 3(16+1)$$

$$= 11.34$$

Since $I = 4$ groups, the sampling distribution of H is approximately chi-square with 4 - 1 = 3 degrees of freedom. From Table G we see the P-value is approximately 0.01. There is strong evidence of a difference in seedling growth between the four groups. As in the one-way ANOVA, nonparametric multiple comparisons or contrasts would be required to explore the group differences further.

The MINITAB software gives the output below when doing the Kruskal-Wallis test. The medians, average ranks (in place of sums of ranks), H statistic and P-value are given. The H statistic with an adjustment for ties in the ranks is also given.

Kruskal-Wallis Test

LEVEL	NOBS	MEDIAN	AVE. RANK
1	4	10.000	12.3
2	4	11.100	12.8
3	4	5.200	4.2
4	4	5.550	4.7
OVERALL	16		8.5

$H = 11.34$ d.f. = 3 p = 0.010
$H = 11.35$ d.f. = 3 p = 0.010 (adjusted for ties)

CHAPTER 15

LOGISTIC REGRESSION

OVERVIEW

In Chapter 8 we studied random variables y which can take on only two values (yes or no, success or failure, live or die, acceptable or not). It is usually convenient to code the two values as 0 (failure) and 1 (success) and let p denote the probability of a 1 (success). If y is observed on n independent trials, the total number of 1's can often be modeled as binomial with n trials and probability of success p. In this chapter, we consider methods that allow us to investigate how y depends on one or more explanatory variables. The methods of simple linear and multiple linear regression do not directly apply since the distribution of y is binomial rather than normal. However, it is possible to use a method, called **logistic regression**, which is similar to simple and multiple linear regression.

We define the **odds** as $p/(1-p)$, the ratio of the probability that the event happens to the probability that the event does not happen. If $\hat{p}$ is the sample proportion, the sample odds are $\hat{p}/(1-\hat{p})$. The **logistic regression model** relates the (natural) log of the odds to the explanatory variables. In the case of a single explanatory variable x, the logistic regression model is

$$\log\left(\frac{p_i}{1 - p_i}\right) = \beta_0 + \beta_1 x_i$$

where the responses y_i, for i = 1,..., n, are n independent binomial random variables with parameters 1 and p_i, i.e., they are independent with distributions $B(1, p_i)$. The **parameters** of this logistic regression model are β_0 (the intercept of the logistic model) and β_1 (the slope of the logistic model). The quantity e^{β_1} is called the **odds ratio.**

The formulas for estimating of the parameters of the logistic regression model are, in general, very complicated, and in practice the estimates are computed using statistical software. Such software typically gives estimates b_0 for β_0 and b_1 for β_1 along with estimates SE_{b_0} and SE_{b_1} for the standard errors of these estimates. A **level C confidence interval for the intercept** β_0 is then determined by the formula

$$b_0 \pm z^* \, SE_{b_0}$$

where z^* is the upper $(1-C)/2$ quantile of the standard normal distribution. Similarly, a **level C confidence interval for the slope** β_1 is determined by the formula

$$b_1 \pm z^* \, SE_{b_1}$$

A **level C confidence interval for the odds ratio** e^{β_1} is obtained by transforming the confidence interval for the slope, yielding the formula

$$\left(e^{b_1 - z^* SE_{b_1}}, \; e^{b_1 + z^* SE_{b_1}} \right)$$

To **test the hypotheses** H_0: $\beta_1 = 0$, and H_a: $\beta_1 \neq 0$ compute the **test statistic**

$$X^2 = \left(\frac{b_1}{SE_{b_1}} \right)^2$$

Since the random variable X^2 has an approximate χ^2 distribution with 1 degree of freedom, the P-value for this test is $P(\chi^2 \geq X^2)$. This is the same as testing the null hypothesis that the odds ratio is 1.

In **multiple logistic regression,** the response variable again has two possible values, but there can be more than one explanatory variable. Multiple logistic regression is analogous to multiple linear regression. Fitting multiple logistic regression models is done, in practice, with statistical software.

GUIDED SOLUTIONS

Exercise 15.1

KEY CONCEPTS - odds, odds ratio

a) You need to compute

$$\hat{p}_{high} = \frac{\text{\# in high blood pressure group who died of cardiovascular disease}}{\text{total number of men in the high blood pressure group}}$$

$$=$$

and then

$$\text{Odds} = \frac{\hat{p}_{high}}{1 - \hat{p}_{high}} =$$

b) Repeat the type calculations you did in (a), now for the men with low blood pressure.

$$\hat{p}_{low} =$$

$$\text{Odds} =$$

c) In this setting, recall that the odds ratio is

$$\text{Odds ratio} = e^{\beta_1} = \frac{\text{Odds for men with high blood pressure}}{\text{Odds for men with low blood pressure}} =$$

Interpret this odds ratio in words.

Exercise 15.3

KEY CONCEPTS - logistic regression, confidence interval for the slope, test that the slope is 0

a) A 95% confidence interval for the slope β_1 is determined by the formula $b_1 \pm z^* \text{SE}_{b_1}$, where z^* is the upper 0.025 quantile of the standard normal distribution.

Look up $z*$ in Table A and then, using the information given in the problem, compute

$$b_1 \pm z* \, SE_{b_1} =$$

b) Recall that the X^2 statistic for testing the null hypothesis that the slope is zero is

$$X^2 = \left(\frac{b_1}{SE_{b_1}} \right)^2$$

and has a χ^2 distribution with 1 degree of freedom. Use the values for b_1 and SE_{b_1} given in the problem to compute this quantity.

$$X^2 = \left(\frac{b_1}{SE_{b_1}} \right)^2 =$$

Now use Table F to find the approximate P-value; i.e.

$$P\text{-value} = P(\chi^2 \ge X^2) =$$

c) Now summarize your results and conclusions. What can you conclude about the probability of death from cardiovascular disease for men with high blood pressure versus that for men with low blood pressure?

Exercise 15.5

KEY CONCEPTS - confidence interval for the odds ratio

a) Recall that a level C confidence interval for the odds ratio e^{β_1} is obtained by transforming the level C confidence interval for the slope, $b_1 \pm z* \, SE_{b_1}$ to yield the formula.

$$\left(e^{b_1 - z* SE_{b_1}}, \; e^{b_1 + z* SE_{b_1}} \right)$$

Referring to Exercise 15.3 (a), what is the 95% confidence interval for the slope? Hence, what is a 95% confidence interval for the odds ratio?

b) What does the interval you found in (a) tell you about the odds that a man with high blood pressure dies from cardiovascular disease relative to the odds that a man with low blood pressure dies from cardiovascular disease?

Exercise 15.13

KEY CONCEPTS - multiple logistic regression

a) You will need access to statistical software that allows you to do multiple logistic regression. SAS, SPSS, Minitab (later versions), and JMP will allow you to do multiple logistic regression. Consult your software's manual for details and explanation of the output. The output should include the value of X^2 (or z) for testing that the coefficients of both the explanatory variables are 0 and will probably also give the P-value of the test. If the P-value is not given, refer to the chi-square table in the text. The appropriate degrees of freedom in this case is 2.

b) Your statistical software should provide estimates of each of the coefficients and the standard errors of these coefficients. 95% confidence intervals can then be constructed using the formula

parameter estimate $\pm z^*$(standard error of parameter estimate).

c) What do you conclude based on your findings in (a) and (b)? Would you use both SATM and SATV to predict HIGPA, just one, or neither?

COMPLETE SOLUTIONS

Exercise 15.1

a) The proportion of men who died from cardiovascular disease in the high blood pressure group is

$$\hat{p}_{high} = \frac{\text{\# in high blood pressure group who died of cardiovascular disease}}{\text{total number of men in the high blood pressure group}}$$

$$= \frac{55}{3338} = 0.0165$$

The odds are

$$\text{Odds} = \frac{\hat{p}_{high}}{1 - \hat{p}_{high}} = \frac{55/3338}{1 - (55/3338)} = \frac{55/3338}{3283/3338} = \frac{55}{3283} = 0.0168.$$

b) The proportion of men who died from cardiovascular disease in the low blood pressure group is

$$\hat{p}_{low} = \frac{\text{\# in low blood pressure group who died of cardiovascular disease}}{\text{total number of men in the low blood pressure group}}$$

$$= \frac{21}{2676} = 0.0078$$

The odds are

$$\text{Odds} = \frac{\hat{p}_{low}}{1 - \hat{p}_{low}} = \frac{21/2676}{1 - (21/2676)} = \frac{21/2676}{2655/2676} = \frac{21}{2655} = 0.0079.$$

c) The odds ratio, with the odds for the high blood pressure group in the numerator is

$$\text{Odds ratio} = \frac{\text{Odds for high blood pressure group}}{\text{Odds for low blood pressure group}} = \frac{55/3283}{21/2655} = 2.1181.$$

This tells us that the odds that, in the study group, a man with high blood pressure dies from cardiovascular disease are slightly more than twice the odds that a man with low blood pressure dies from cardiovascular disease.

Exercise 15.3

a) We are given in the problem that $b_1 = 0.7505$ and $SE_{b_1} = 0.2578$. For a 95% confidence interval, from Table A we have $z^* = 1.96$, and so

$$b_1 \pm z^* SE_{b_1} = 0.7505 \pm 1.96(0.2578) = 0.7505 \pm 0.5052$$

or equivalently, the interval (0.2453, 1.2557).

b) We compute

$$X^2 = \left(\frac{b_1}{SE_{b_1}}\right)^2 = \left(\frac{0.7505}{0.2578}\right)^2 = (2.9112)^2 = 8.4751$$

Using Table G for the χ^2 distribution with 1 degree of freedom, we estimate the P-value to be

$$P\text{-value} = P(\chi^2 \geq X^2) = P(\chi^2 \geq 8.4751)$$

which is between 0.005 and 0.0025.

c) There is strong evidence that the slope parameter in the logistic regression model is different from 0 (or equivalently, that the odds ratio is different from 1). In addition, we are 95% confident that the slope parameter has a value between 0.2453 and 1.2557. Although the problem does not say so, these results, along with those of Exercise 15.1, suggest that the explanatory variable must be an indicator variable with value 1 for men with high blood pressure and value 0 for men with low blood pressure. Since a positive slope parameter implies that the probability of death from cardiovascular disease increases as the explanatory variable increases, we may conclude that the data provide strong evidence that the probability of death from cardiovascular disease is higher for men with high blood pressure than for men with low blood pressure.

Exercise 15.5

a) We saw in (a) of Exercise 15.3 that a 95% confidence interval for the slope in the logistic regression model is (0.2453, 1.2557). Hence, a 95% confidence interval for the odds ratio is

$$\left(e^{b_1 - z^* SE_{b_1}}, e^{b_1 + z^* SE_{b_1}}\right) = (e^{0.2453}, e^{1.2557}) = (1.2780, 3.5103)$$

b) We are 95% confident that the odds that a man with high blood pressure died from cardiovascular disease is between 1.2780 and 3.5103 times higher than the odds that a man with low blood pressure died from cardiovascular disease.

Exercise 15.13

a) The results of running a multiple logistic regression model to predict HIGPA from SATM and SATV are given below. We used JMP to perform the analysis.

Response: HIGPA
Converged by Gradient
Whole-Model Test

Model	−LogLikelihood	DF	ChiSquare	Prob>ChiSq
Difference	7.11016	2	14.22031	0.0008
Full	140.55964			
Reduced	147.66980			

The information regarding the hypothesis test that the coefficients for both explanatory variables are zero is found in the row labeled "Difference" in the table above. The entry under the column labeled "ChiSquare" gives the value of the test statistic X^2. This test statistics has value $X^2 = 14.22031$ and has 2 degrees of freedom (as indicated in the column labeled "DF"). The P-value of the test is given in the column labeled "Prob>ChiSq" and is 0.0008. We would reject the null hypothesis that the coefficients for both explanatory variables are zero and conclude that at least one of the explanatory variables is helpful for predicting the odds that HIGPA is 1, or equivalently, that GPA ≥ 3.00.

b) The information for the individual parameter estimates is summarized below.

Response: HIGPA
Converged by Gradient
Parameter Estimates

Term	Estimate	Std Error	ChiSquare	Prob>ChiSq
Intercept	4.54290625	1.1617665	15.29	<.0001
SATM	−0.00369	0.0019136	3.72	0.0538
SATV	−0.003527	0.0017518	4.05	0.0441

The estimates of the coefficients for SATM and SATV are given in the table above in the column labeled "Estimate." The standard errors of theses estimates are given in the column labeled "Std Error." Recall that the general formula for a 95% confidence interval would be

parameter estimate $\pm z^*$(standard error of parameter estimate)

with $z^* = 1.96$ for a 95% confidence interval. Using the information in the table, we find the following.

95% CI for coefficient of SATM: $-0.00369 \pm 1.96(0.0019136)$
$$= -0.00369 \pm 0.00375$$
$$= (-0.00744, +0.00006).$$

95% CI for coefficient of SATV: $-0.003527 \pm 1.96(0.0017518)$
$$= -0.003527 \pm 0.003434$$
$$= (-0.006961, -0.000093).$$

c) For predicting the odds that HIGPA is 1, or equivalently, the odds that GPA $\geq$ 3.00, the results of (b) show that the 95% confidence interval for the coefficient of SATV does not contain 0; hence, the coefficient would be declared significantly different from 0 at the 0.05 level when SATV is added to a model containing SATM. This suggests that SATV improves prediction when added to a model already containing SATM. However, the 95% confidence interval for the coefficient of SATM just barely contains 0, and we would not declare the coefficient to be significantly different from 0 at the 0.05 level when added to a model containing SATV. Since 0 is very near the upper limit of the 95% confidence interval for SATM, we might nevertheless be inclined to include SATM in the model. This is further supported by the fact that the P-value for the test that both coefficients are zero is quite small. The data suggest that using both SATM and SATV as predictors is reasonable. If only a single predictor is to be used, the data suggest that SATV should be used rather than SATM.